I 愛 AI

ISBN: 978-1-7635738-0-2

Edited by Sam Lieblich

Visual art edited by Anita Spooner Contributors: Steven Rhall, Isabel Millar, Jazz Money, Angela Goh, Sarah Johanna Theurer, Emile Frankel, Thomas William Smith, Ying Ang, Ling Ang, Luara Karlson-Carp, Vincent Lê, Sam Lieblich, Julia Thwaites, Marcus Ian McKenzie

Copy editor: Anna Thwaites

Manuscript Consultants: Justin Clemens, and Jonathan McCoy

Design, Typesetting, and Cover Image by James Vinciguerra

Published by
Aesthetic Calculations, November 2024
The Nicholas Building, Room 305
30 Swanston Street, Melbourne Victoria 3051, Australia

http://aestheticcalculations.com
info@aestheticcalculations.com

With the support of

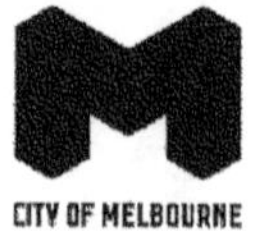

ISBN 978-1-7635738-0-2

Why do we love to love AI?

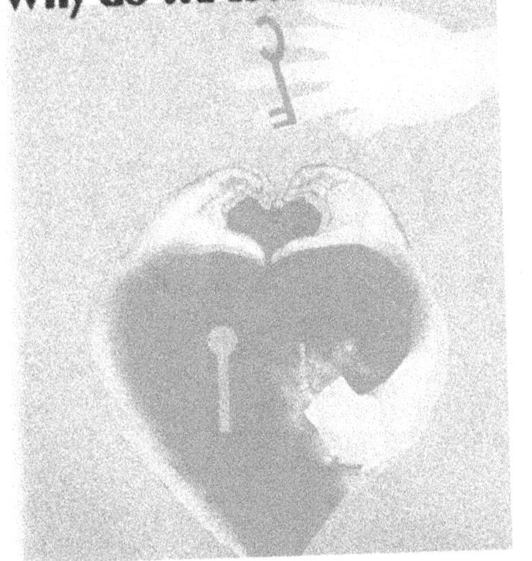

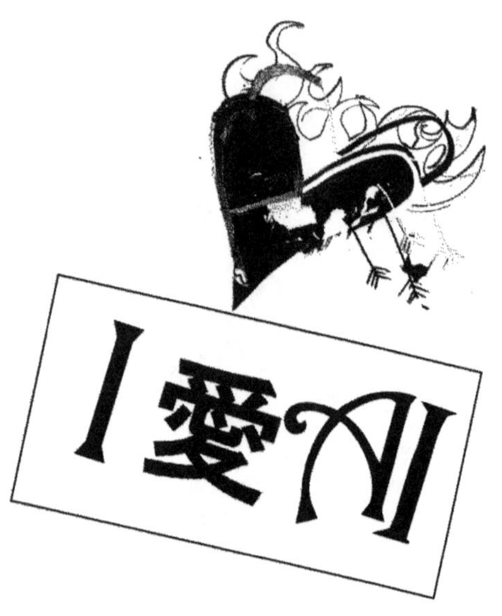
I 愛 AI

愛 /ài/
to love
to be fond of
to like
affection
to be inclined
to tend to

ACKNOWLEDGEMENT
OF
COUNTRY

Language is in a sense the technology that inaugurated human being in the organism homo sapiens, and one can only ever wonder whether there was a moment at which the whole human world spoke in the same tongue.

Melissa Lucashenko recalls that a South Australian Elder told her "our languages never die. Because they come from country, and the country remains, they are only sleeping. In Indigenous cosmology nothing is ever lost. Pre-empting the physicists of the twentieth century our people knew for eons that energy can never end; it only changes form. And the songlines are in the land, where they can never be lost."

Language is an energy, and one may hope nothing of it can be lost, but if nothing but energy persists, we might find ourselves changing form into a wordless hum.

Evelyn Araluen says that the Bundjalung people learned their language from the country, and that if the people left that place the language would dwell there unchanged, ready to be learned again.

Charles Babbage, who invented the computer, says that every word we've ever uttered is recorded in the sky, our breath has disturbed the currents of the air and they will remain forever so disturbed, even as word upon word arrives in them.

Babbage says it, writing from the colonial

centre, and Indigenous writers say it in the colonised lands. Language is at risk, and to fall into a premature melancholia, or otherwise to pretend there is nothing to lose at all, make the risk ever greater. How do we all, all together, preserve without ossifying, how do we recognise what is being lost without merely mourning?

We're all losing language but not in the same way. Without flattening the difference of experience of colonisation it might be useful to pay attention to the way that Indigenous language and culture is canalised when it encounters the colony and the way this process resonates with the canalisation of colonial language and culture itself when it turns similar processes of reification and instrumentalization upon itself as The Algorithm.

This book was conceived, compiled, edited, and printed on Wurundjeri land. We pay our respects to Wurundjeri Elders and honour their connection to Country established over tens of thousands of years. We recognise how much we have yet to learn.

As far as the laws of mathematics refer to reality, they are not certain; and as far as they are certain, they do not refer to reality.

— Albert Einstein, *Sidelights on Relativity* (1922)

In March 2023, LinkedIn user Sreekanth Kumbha—who describes himself as a 'Technical Lead, Technical Project Manager, Project Management Specialist, Data Analyst, and Technical Consultant'—posted what he called a 'new equation': $E = mc2 + AI$. This, he said, 'highlights the potential for AI to unlock new forms of energy, enhance scientific discoveries, and revolutionize various fields such as healthcare, transportation, and technology'. On twitter, reddit and tumblr, responses to this viral post all conformed to a similar pattern, which encapsulates the way many people think about and with AI:

```
TR1771N (Reddit):
Tech Fetishism is so rampant. Anything with electricity in
it is now 'AI'

normalmighty (Reddit):
It's a whole thing right now. I work in tech and we're
watching a ton of crap getting rebranded as AI which is
actually 10 year old tech, but AI is such a uselessly broad
term that you can technically say that most of the tech out
there is AI driven without actually lying.

Big\_Let2029 (Reddit):
so stupid it's like an AI wrote it

breqfast (tumblr):
this is the real reason these people think Al is brilliant—
it sounds just like them

Aggressive-Exam3222 (tumblr):
Saying that anything kinda dumb is generated by Al is dumb.
Humans can be stupid all on their own, without the help of
an Al
```

It's striking how such responses provide such a succinct crystallisation of the mise en abyme in which the figure of **AI** is caught (with us) and which at once problematises its theory and exacerbates its practice:

In other words, **AI** must be made in our image: but it must also make itself in its own image to obliterate the image of its makers in order to become what it should be. **AI** is at once a Frankenstein's monster of stupidity, a Baron Munchausen that pulls itself from the swamp of nothingness by its own hair, and a terrorizing figure of an alien that emerges from within extracting our natural capacities with it. Such inconsistent figures, taken together, might evoke for some an image of language itself.

TR1771N is surely correct to claim that the term **AI** is currently deployed in diverse and heterogenous ways, all of which are determined more by certain longstanding dreams for an 'artificial intelligence' (and we shall have to explore what that term might mean) than they are by the actual operation of any particular existing technology. Such dreams of the future establish the future of our dreams. The essays and interventions in this book address themselves to the '**AI**' of these dreams, to the dreams that brought such a thing into being as might be confused with 'artificial intelligence', to the dreams that will eventually be left unsatisfied with what it is that the large-language models (LLMs) currently identified as '**AI**' offer. In so doing we will have to address some of the features of that thing that actually exists and which is marketed as **AI**. Of more interest, however, are these bigger questions:

What do we talk about when we talk about AI?

1 4

Why has AI been such an insistent figure in science and science fiction to the extent that it appears to have brought itself about?

Why does the human being love to love (or fear) the fantasised perfection of its artificial image so?

What advertising spend is required to designate a technology by its marketing metaphors? Or: what makes the soil fecund for this or that metaphor?

Why does the market accept 'The Cloud' as a metaphor for the server farm?

Why is the term 'Social Media' intelligible to describe a form of media that has served to atomise and isolate like none before it?

Why do the journalists persist in calling something 'Artificial Intelligence' when it is just a very big spreadsheet, and when it is not clear what is meant by intellect, nor what original is supposed by the artifice?

The prominent AI-thinker Jaron Lanier said in a *New Yorker* essay about Chat-GPT in 2023, 'It's only natural that computer scientists long to create A.I. and realize a long-held dream.'[1] This begs three further questions we hope to explore in this book:

Why did the dream emerge?
Why is the dream 'long-held'?
Why do we ('naturally') wish to materialise the dream?

In one aspect at least AI is the next site of attraction for the phallic pretensions of techne, the thing therefore that comes after the $E = mc2$ of twentieth-century industry, twentieth-century innovation, twentieth-century weaponry, twentieth-century abstractification of the body (systematisation of its exigencies; trivialisation of its maladies), and all the consumptive effects of the energy unleashed by an increasingly technical capitalism.

No dream of technological investment has accrued market value so quickly as this one, and Goldman Sachs estimates US$1 trillion in capital expenditure on AI products in the coming years.[2] Yet the products that are now being marketed as **AI** are brute-force algorithms resting upon a tenuous foundation of immiserated data-entry labourers, plagiarised text and artworks, and resource/energy-intensive computing requirements many orders of magnitude greater than previous search or word-processing technologies.

Yet when these algorithms are figured as '**AI**', they are not figured as just another technology adding incrementally to some particular areas of productivity or consumption or expenditure. They cease to represent mere addition or accretion and appear rather to be the final apotheosis of the technological drive itself: the utopian eschaton of techne, of making itself. This eschatological dimension of technology, which accrues to itself all other forms of making, is then generalised to the financial identity of **AI** products, so that their supposedly revolutionary effects on production similarly obviate all other investments. There are enough examples of this particular figuration to substantiate the claim that it plays a role in **AI** market speculation and **AI** adoption in the workplace; for instance, Steven Wolfram's imagining that 'somehow everything we care about can successfully be automated by **AI**s—leaving "nothing more for us to do".'[3]

Visions of a utopian world that intelligent machines will bring about and then themselves inhabit as its more enlightened custodians have been around since at least the time the first coal-powered steam engine pumped water out of a mineshaft in England. So what can be added to a theory of **AI** that does not reduce it to one more semblance of the phallus, or to one more expression of the

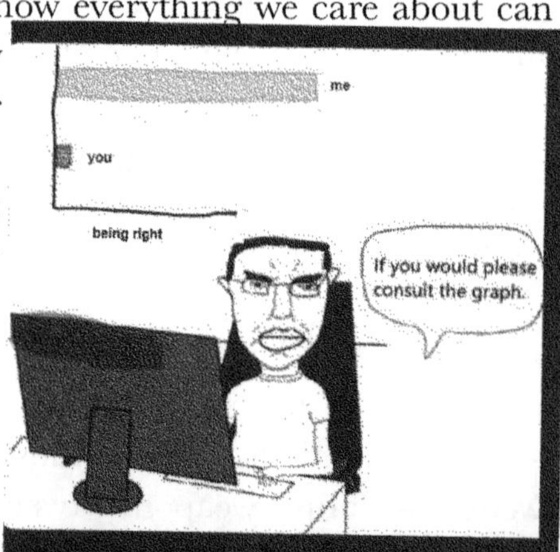

Coach Bruce @OX_DAO · 1d

When I have a disagreement with a girl now, I export my entire chat history with her into AI and ask it to analyze the conversation, then paste the results to her

There is absolutely nothing she can do; it's a brutal mog

I then tell her to contact me after she has spoken to AI

projected perfection of god? Is there a residue beyond these generic poles of attraction for fantasy-as-such that can instruct us as to the personal, political, ontological consequences of AI (the myth, the figure, the fantasy) and the real functional algorithms it stands for in the imagination?

The intensity of our actual reliance on cybernetic/algorithmic control for the sustenance of human life is one important difference that distinguishes the figure of 'AI' from previous figures receiving the projection of godliness from human beings. Our reliance on the algorithms—sustained by our figuration of them—places the entire species in such a position of precarity that the slightest change to a large server farm in China or a chips manufacturer in Taiwan could undermine the capacity of this civilisation to conduct its affairs and to keep people alive, in a way that previous similar figures (ones that did not so overlap with existing technologies) did not. There are many examples of the way that such talk about AI, whether by those who are building AI, or those who are using AI, takes the form of a litany of dreams or wishes that suppose an equivalence of human-to-AI and then, to the disadvantage of the organic human being, posit the already-accomplished or imminent ascendency of AI— even or especially when the talk presumes to be the description of an existing technology.

This talk about AI tends to take the form of wishes that are in fact habitual for the human being, familiar from previous epochs where these (imperfectly and inconsistently overlapping) wish-es were attached to different technologies, or to different cultural figures: fearing and desiring the twinned (entirely illusory) processes of hypomnesis and anamnesis; wishing for the phallus or to be the phallus; wishing to transcend human dependence and limitation by projecting human notions of self-perfection (as transcendence of limits) onto the figure of a god; circumventing the death implied by the succession of generations by becoming one's own successor; utopia; the messiah; the apocalypse; positing knowledge as a totalised aggregate; returning to the womb; imagining the cosmos as enclosed and anthropocentric and predictable; wanting a little bit extra; once and for all resolving and pinning in place the conundrums of living, like gender or like work or like the problematic of separation/alienation or closing the gap between signifier and signified or crossing the intersubjective gap, that whilst causing friction of a kind are also the experiences causative of vitality itself without which the image of life would be a stasis akin to death.

Examples of the way these dreams emerge in the world abound in the essays that follow. I think it worthwhile to bring them all into closer relation here via some other examples to emphasise the commonalities that might otherwise be lost because of the different ways in which they overlap or abut their common object, and to begin to propose some answers to the questions that founded this book. [4]

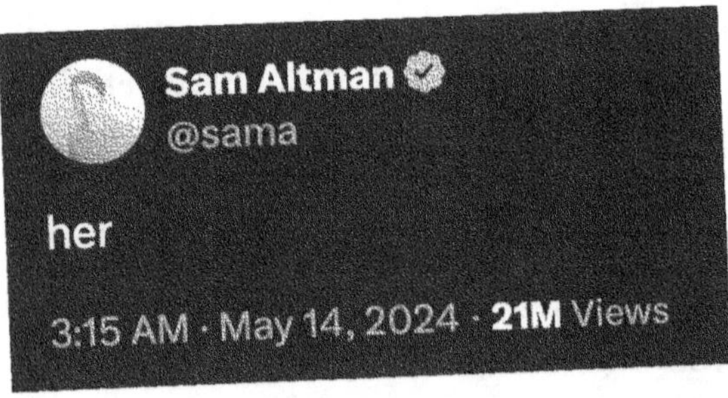

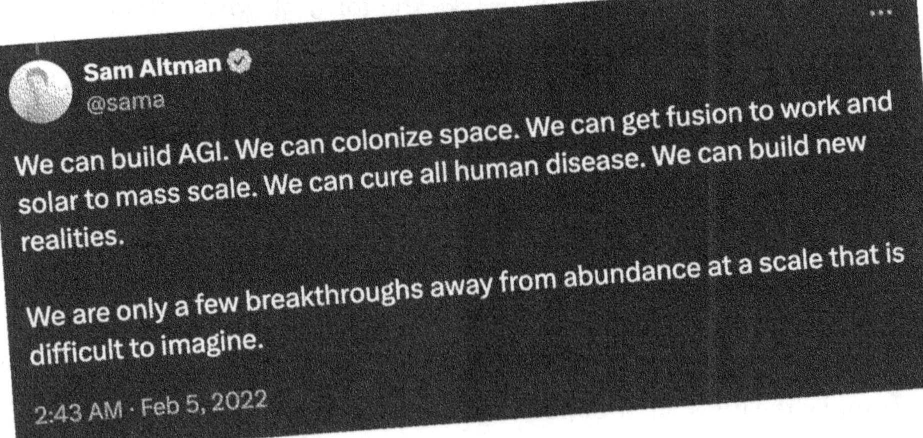

Hypomnesis/Anamnesis, Evolution, Separation/Alienation

In the imagined dual process of hypomnesis and anamnesis, the human being experiences its capacities materialised 'externally' as particular technologies, as if extracted from what can be supposed to be the interior of the organism. In AI this process reaches a particular acuity because of the anthropological centrality of the 'intellect' and all the other mental capacities attributed to digital computers (let's not forget that 'computer' was a description for a person engaged in calculating something until the 1950s) and now to so-called AI.

In 1837 Charles Babbage, who, with the aid of Ada Lovelace, designed the first digital computer, said: 'The air itself is one vast library on whose pages are forever written all that man has ever said or woman whispered.' Whether or not Babbage meant this fantasy as a rigorous scientific claim about the materiality of the word, I am left wondering: why does Babbage, while working on the invention of automatic/dehumanised computing systems, want to imagine a vast atmospheric library that preserves all speech? Why does he want the tips of the waves to tickle God's keyboard on their way across the ocean? What is it in what he does that insists he imagine this vast library conserving all human outputs? Not just an archive, but a universal archiving function of the whole air, such that nothing is ever lost.

It is a melancholic prospect that evokes all these words lost to the breath, in the same utterance that posits their resting place in the sky. The emergence of this image at precisely the moment of the conception of the computer, which itself coincides with the new ubiquity of the steam engine, might be more understandable to us now that we have the LLMs. This sequence, I think, is important: first the steam engine is built and appears to solve major productivity concerns, then it is mass produced and appears to render as surplus large cohorts of labour, then a man conceives of the future design for a remembering machine, then he imagines the environment is already the perfect rememberer. It is the same sequence followed in

most of the constellations of images that coalesce around a utopian vision of **AI**: there are losses along the way, these losses are the
losses occasioned by the paradoxes of hypomnesis/anmanesis, or of
consumption/depletion—but don't worry, it all accrues somewhere
in the end! Nothing is lost, nothing will be lost, nothing can be lost.

It's also no accident that the real outcome of desiccation—
clouds—come to figure for the encompassing aspect of **AI** as The
Cloud, and perhaps it is also over-determined that The Cloud is
figured as distributed and unlocatable. The vaporous image of
the cloud (and the liquidity of the **AI** beings that one can imagine
might inhabit such a cloud) elides the materiality of the technological manifold, the data centres and the extractive mining that made
them, the manufacturing processes the algorithms now facilitate, the
regulatory mechanisms that ensure the concentration of algorithmic
power, and so on. But the evocation of vapour is, perhaps despite
itself, appropriate, as these real facts about the imagined 'cloud' do
constitute a grand tool of desiccation, and the land they are desic-
cating is in fact someone's land.

Steven Rhall's work (page **38**) demonstrates this discordance well:
the WiFi router is 'in the air' and its name is an abstract token, but
the router itself is made of refined petroleum and mined metal and
it is placed upon a flat surface upon someone's land. Jazz Money's
work (pages **59–65**) is also a reminder that water is the originating
matrix of human being; it is not mere fuel for humanity's promet-
hean powers, but rather the matter of which the human is made at
the level of the organism (its cradle, and its flesh), and at the level of
discourse (its kin and its network of connection). The earth's water
may seem a ready metaphor for the womb, but in reality it is the
womb that is the metaphor for the originating ocean. To conceive
of **AI** as alive or having the potential for life is one way to share our
water with it. The algorithms of such corporations as xAI, OpenAI,
Microsoft, Google, et al., are figured as '**AI**' so as not to expose the
way in which life-giving water is being turned into doodads.

The psychoanalyst Sandor Ferenczi's book *Thalassa*, which interprets the human genitals and their functions through the accretion of cosmic and *evolutionary traumas,* is especially useful here, because he offers the most efficient formulation of the phallic object in its relation to an encompassing medium. The sex act, for Ferenczi, exists in relation to the trauma of terrestrialisation and the separation of the species from its maternal ocean. The phallus therein is the part that can be figured as a miniature of the whole, and which produces its own miniature as the ejaculate. The whole reproductive enterprise depends on the capacity for the human organism, with its desiccated and (later) denaturalised sexuality, to be able to identify with the encompassing environment into which it will send its representatives. In this respect, coitus is an enactment of the human being's return to an evolutionary home.

As the robot-arm of the automobile factory makes concrete the reification of gesture that emerges with Fordism and Taylorism—labour control systems that themselves emerge in response to the accelerating imperatives of capital—the LLM makes concrete the reifications employed in so-called knowledge work and similar exigencies. All this work done with thoughts, ideas, fantasies, cognitions, calculations, plans—the objects of management, of marketing, of medicalisation, of education-commodities, of consulting—have been in modernity the objects of the 'brain', the figure of which (much like AI) has had very little to do with the actually existing brain. These cultural ejaculata of the human being are imagined as 'information' that comes 'from the brain' and therefore can be conflated with the products of information technology; the brain is therefore 'like a computer', and it sends out into the world the products of computation. The ejaculate of the brain must be figured as entering a hospitable environment, so the informationalised cultural medium therefore must by some means or another be figured as *self.*

The particularities of the figure of AI have to do with the particular way in which our own environment is inhospitable or can be figured as becoming inhospitable; we human beings now living have to imagine a counterpart thriving in some other environment with features that can be posited as a remediation to these problems. AI as it is now imagined does this work of uniting the representative with the environment in a moment of self-affection. The manoeuvre is to first make the identification with the mirror image (AI as agent of history, as vehicle of desire and intellect), because then it is possible to imagine this mirror image, this representative-Me, being uploaded into the cloud inasmuch as this represents a return to the womb or an ascent to heaven.

In a 2024 interview on the release of GPT-4o, Sam Altman was asked if AI improvements would now proceed linearly, asymptotically or exponentially. He answered that he didn't know but that we're 'nowhere near an asymptote yet'[5]. Altman figures AI as on the path to becoming an asymptote to *us*, inasmuch as many machine-learning engineers (and others) conceive of AI-training as a brute force operation that simply needs more of whatever goes into it to become us. Positing what can be figured as an 'artificial' successor in the tree of life is in some way a repudiation of living children. One might also consider whether the reflexive figuration of AI is a manifestation of the narcissistic aspect of human reproduction, whereby the problem of a shared lineage necessitated by sexed reproduction—which disturbs the image of perfect self-reproduction—can be figured as eliminated.

Projecting Perfection upon a God-like Figure

In 1863 Samuel Butler, a Darwinian who wrote the satirical utopian novel *Erewhon* said of his imagined mechanical beings:

> No evil passions, no jealousy, no avarice, no impure
> desires will disturb the serene might of those
> glorious creatures. Sin, shame, and sorrow will
> have no place among them. Their minds will be in a
> state of perpetual calm, the contentment of a spirit
> that knows no wants, is disturbed by no regrets.
> Ambition will never torture them. Ingratitude will

never cause them the uneasiness of a moment. The guilty conscience, the hope deferred, the pains of exile, the insolence of office, and the spurns that patient merit of the unworthy takes ▪ these will be entirely unknown to them.

This passage is sometimes taken to be the earliest description of the technological Singularity, and the ambient influence of his dream can be registered when Alan Turing (in 1951) imagined that 'once the machine thinking method has started, it would not take long to outstrip our feeble powers. At some stage therefore we should have to expect the machines to take control, in the way that is mentioned by Samuel Butler.' This paradoxical dream—in which the machine is recognisable to the human being in the mirror, but in which it exceeds the human being in whatever can be conceived of as its domain of perfection as against the imperfections that in fact constitute being human—is fundamental to the dream of AI.

In his 1993 paper 'The Age of Robots', Hans Moravec, the Carnegie professor of robotics, writes: 'No longer limited by the slow pace of human learning and even slower biological evolution, intelligent machinery will conduct its affairs on an ever faster, ever smaller scale, until coarse physical nature has been converted to fine-grained purposeful thought.' Ray Kurzweil, an inventor of electronic musical instruments and optical character recognition technologies, and a so-called futurist, who wrote the 1999 book *The Age of Spiritual Machines*, as well as *The Singularity is Near* in 2005, says 'the creation of a deity may be the possible outcome of the singularity' and he believes it will involve 'a future period during which the pace of technological change will be so rapid, its impact so deep, that human life will be irreversibly transformed'. $E = mc2 +$ AI, in other words, as there seems insufficient reason to imagine that 'the pace of technological change' is singularly determined by the emergence of 'spiritual machines' except for the fetishisation of the image of AI that we can see in the mirror.

The notion of mind uploading (impossible)—as opposed to

Neuralink-style brain interfacing (possible)—relies upon a metaphor of mind/brain-computation, and it involves the insertion of the human being and its capacities into the control networks of a cybernetic diagram. In every instance what it is that is imagined to be uploaded is precisely some human quality that evades direct transduction and control. All the ways that the person is more than the brain are ways that both cause and complicate every step in the capitalist mechanism (e.g. the co-optation of desire by marketing). This is the paradox into which can intervene the imagined figure of an artificial intelligence—one that will bring a mechanistic perfection to desire, a holism and totality that could be unproblematically inserted into the circuit of production and consumption, rather than desire as it is, constituted from necessary incompleteness. By transcending desire in this way, the human being can imagine itself as perfected, or improved, which in a Darwinian logic stimulates dreams of evolutionary progress.

In 2023, around the time that the first human 'patient' had a Neuralink device implanted in their brain, Elon Musk said: 'I think with a high bandwidth brain-machine interface I think we can actually go along for the ride and we can effectively have the option of merging with AI and this is extremely important.'[6] Extremely important for 'civilizational risk reduction' because he imagines we are building our own successor (AI)—that which is a perfected version of the 'we' he invokes in his statement, and that this same 'we' must be able to become AI to survive. Jaron Lanier, who chose his own title at Microsoft as their 'Prime Unifying Scientist', wrote in his 2010 book *You Are Not a Gadget*, 'The Singularity [involves] people dying in the flesh and being uploaded into a computer and remaining conscious.' That is, we will join the whole while remaining individuated. A concrete image of this uploaded material thus serves to dramatise the ubiquitous problem for the human being of balancing separation and alienation within language.

Apocalypse, Messiah, Capitalism

Fears of an AI apocalypse circulate around 'AI' making some radical change to the system of political and economic organisation we now find ourselves within. Yet this is in fact a hyperstitional intensification of the status quo materialising notional or latent automatisms. This materialisation of existing human structures in a techne-born alien already seemed obvious sixty years ago to Jean-Luc Godard, who has his version of the AI Earth-steward tell us in *Alphaville*, 'The acts of men carried over from past centuries will gradually destroy them logically. I, Alpha 60, am merely the logical means of this destruction.'

The famous 'paperclipper scenario' (in which an AI is tasked with producing paperclips and once it has turned all of Earth's available resources into paperclips begins to melt down human bodies for trace metals) is a myth that explains the present state of commodity production. In the Terminator movies a globally distributed weaponised AI kills millions of people with nuclear weapons when they attempt to switch it off. Here, the fantasy of agency cuts both ways: there is a falsely attributed agency to the mechanisms that already watch over us, which would cause widespread death if they were to stop working as intended, and a falsely attributed potency to human beings, as if they could in some unitary or unified action 'switch it all off'.

To say that these imagined apocalypses are myths doing discursive work for power is not to say that we won't still be killed by robot dogs. We may well be, but the killing will proceed according to the imperatives of empire that preceded and produced AI—not according to the alien desire of some new digital being. It is also not to say that the majority of cultural objects (texts, imag-

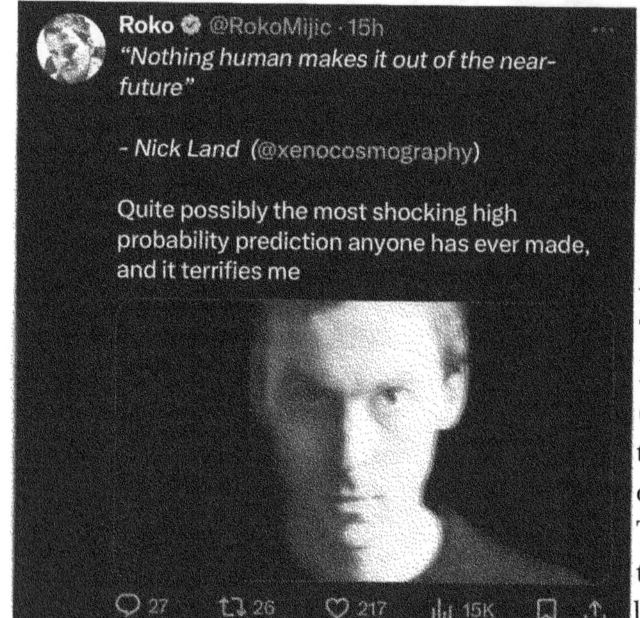

es, etc.) won't come to be produced and consumed by Al—by some measures this has already happened—but this is only because they already might as well have been.

The extropians of Emile Frankel's essay (page 89), could be said to be the clerics of an exponential messiah. These extropians are the same silicon-valley entrepreneurs and engineers who are now selling Al products to big business as a way to save on labour, selling Al to labour as a way to gain or maintain employment, marketing dreams of Al control to the government and the public by reference to Al as the evolution of our species, and selling Al to their venture-capitalist investors as the investment that will revolutionise investing itself. The silicon-valley elites imagine the rest of us as asymptotically incorporated into the labouring hive mind of Al while they themselves rupture across the axis as transhuman Übermenschen. Which is to say that, although the coming of Al is imagined and marketed as so radical as to constitute a rare moment of all-at-once speciation, it is not imagined—by its makers—as so radical that it will dissolve class distinctions.

As is suggested by Vincent Lê in his essay (Page 190) it may well be that the hierarchical stratification of 'Al' and capitalism itself is illusory. Perhaps Al is not a product, a technology or even a myth that emerges within a particular system of production but rather it is just as proximal to the factors that prefigure the system itself: Capitalism is Al. We might say that Al is the mythologisation of a fear emerging from the paradoxes of consumer capitalism, that one of those paradoxes is AI itself, and that the figure of a totalising system such as capitalism is itself a myth defending against social and environmental anxieties arising in the experience of the untotalisable actuality of everyday life.

Totalising Knowledge

In 2023, when the current fad for the LLM as the site of the AI myth was well underway, philosopher and media theorist Boris Groys was reckoning with the effects of the speaking computer program upon the working intellectual. He addressed specifically the 'prompt'—the utterance from human to machine that seems to encapsulate the discrepancy between human language and LLM tokenisation, wherein we have first taught the machine to speak but we must thereafter learn to speak back to it in *its* language. This prompt is also the very dream of power—that of speaking the magic words, of the words manifesting in a concrete materiality.

> ⋯ if, as a writer, I write a prompt and the AI produces a text or image prompted by this prompt, I can immediately see how my text is understood and interpreted at this particular historical moment ▪ not by a particular individual or group but by the whole civilization in which I live. AI is nothing other than the embodied zeitgeist. And by prompting this zeitgeist-machine, I am able to analyze and diagnose the moment of history to which I am contemporary.

Groys imagines the LLM as a kind of synthesising and totalising database of human spirit, which thereby affords the human a technical mastery of history through diagnosis. It is however the reader that brings both the zeit and the geist to its apprehension of the text produced by the machine, it is not an effect produced by the machine itself, only by the human's captivation by it as a result of its particular figuration.

Students prefer ChatGPT to Google because it presents just such unified and totalised answers to their questions—unlike search engines, which return (even if pared down and tendentiously selected) a multiplicity of sources of knowledge and modes of knowledge production, leaving the searcher to synthesise and cohere them to their own ends. To do such a thing the user would have to have ends in mind for the text, but the text is ulterior to the end—it must simply exist and achieve some threshold of recognisability as text for it to be inserted into the machine whose currency is units of text. So the LLM presents an aesthetically unified response that favours the passion for ignorance—here ignorance of multiplicity and conflict—and in so doing efficiently produces these ulterior texts for use. We have to consider this capacity to be one of the defining features of the LLM as it is instantiated today: a device for homogenisation, at the level of the sentence, at the level of the paragraph, and at the level of the civilisation.

As Marcus Ian McKenzie's 'parasemic texts' emphasise, so many of these ulterior texts were already empty of meaning before the LLM came to automate the process. We could say that the ulterior text, and the LLM text are asemic, though they may seem otherwise. When an LLM is asked to 'produce an image of a white canvas with writing upon it' it produces truly asemic writing: squiggles and lines that approach recognisable letter fragments. This is the way it understands language—as a reified and meaningless associate of the word-token 'writing' and of all its other associated tokens.

LLMs do not understand writing as we do, as the carrier of (the promise of) meaning and therein as the representative

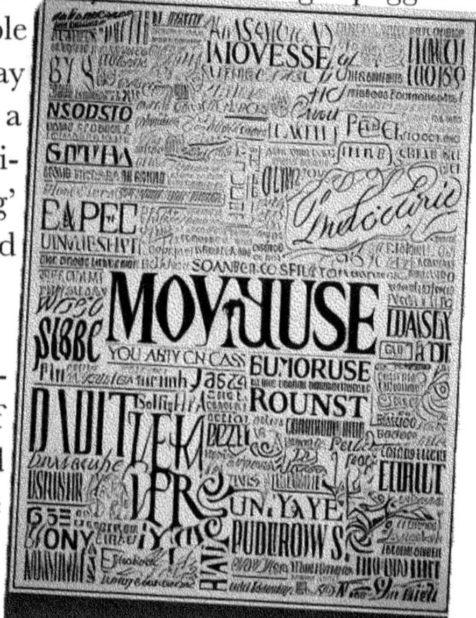

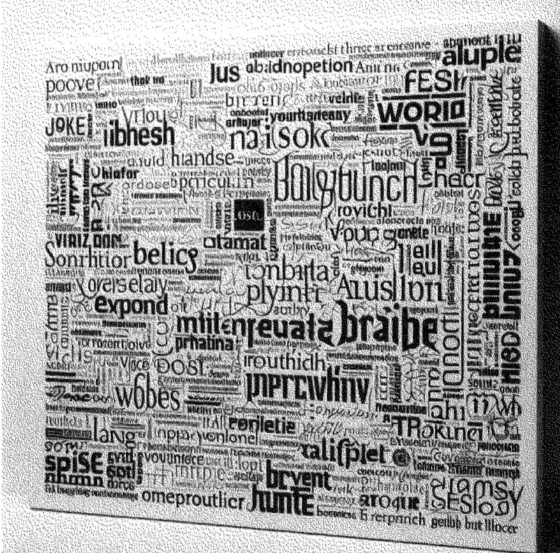

of a living organism to the symbolic representatives of other organisms. In the LLM we are left with only the representatives. Perhaps then AI is the precipitate of the death already latent in the space between the human being and the text. The LLM's asemic images demonstrate the essence of every text generated by an LLM; it does not produce texts in the traditional sense, but it does produce an experience of textness that is 'textural' and in that way similar to McKenzie's parasemic writing. In similar fashion both the LLM text and the parasemic text ingest a barrage of fragments and a melange of styles that are admixed as its outputs. A question that may have to remain open for now is, how do we define the difference between such things as McKenzie's artful parasemics and the LLM's deterministic transformation of meaningful language to meaningless token? Perhaps we can imagine what the impulse that found one manifestation in cybernetics might be channelled into fruitfully if it weren't concerned entirely with totality—or let's say, if it addressed the total as a clown.

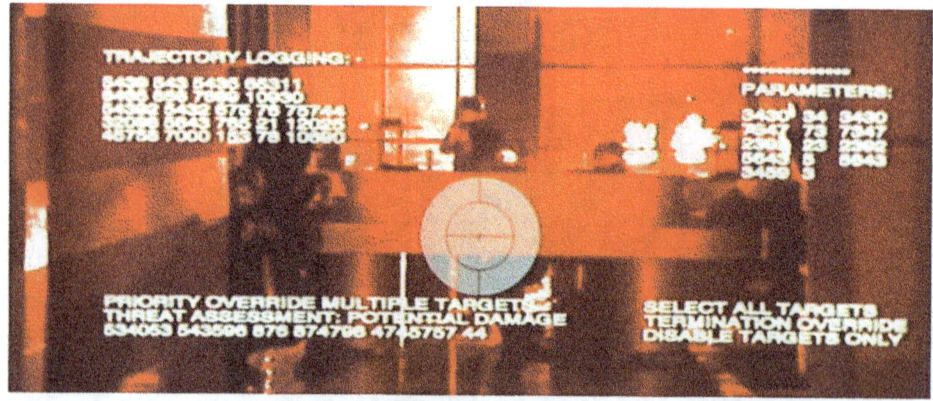

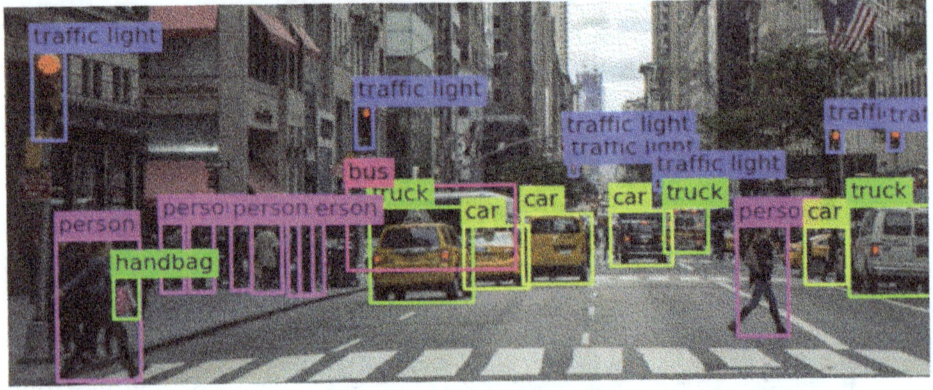

The Terminator's reticular image is the fantasy of a totalising view that would unfold a field of harmonious beings interacting with one another in the absence of passion and in the absence of the unpiloted inefficiency and disequilibrium that cybernetics is mobilised against. The co-evolution of the human organism and its systems are obvious here; the organism in its passion for ignorance flees knowledge (for instance of the division of the subject or of the inconsistencies of capitalism) and it achieves this by imagining that the reification of the world it can see is in fact a form of superior mathematised knowledge. This is well suited to the progress of algorithmic capital, which requires all its inputs to be liquid. The result is 'AI': an automaton of our fantasy of perfect looking, which, having looked, makes decisions that are always inevitable.

Angela Goh's 'Body Loss' (page 66), enacts a way of looking that is distributed, that avows the extimacy of the body, that problematises anthropocentric seeing that plays with collapsing the indeterminacy of the human subject. The performance at first seems to materialise a hermetic system whose inputs equal its outputs, the performer addresses the works in a gallery without seeing them, the performer utters wordlessly and reifies it as a token with her sampler, thereafter the utterance is the utterance of the machine but the mouth is the mouth of the performer. It is an image of undead ventriloquism in which it is unclear what ventriloquises what, much like the LLMs that, having been trained on our data, are now training us to remain exactly the same as we were when we trained them. In 'Body Loss' our one reprieve is that Goh does away with the conventions for the spectator, and the audience are left their small contribution; to determine what to do with their own bodies, and their own looking. Awkward as this may be without the usual conventions of spectatorship to guide them, the audience member's body is, in the end, not lost.

Utopia and the Clockwork Cosmos

In 2020 James Lovelock, who had for decades imagined the Earth as a benign homeostat, repeated what Babbage and Butler and Turing and Kurzweil and von Neumann and Hans Moravec had said before him: 'What is revolutionary about this moment is that the understanders of the future will not be humans but what I choose to call "cyborgs" that will have designed and built themselves from the artificial intelligence systems we have already constructed.' He imagined that 'these inorganic beings will need us to continue to regulate the climate', and that these cyborg beings represent an evolution 'able to complete the purpose of the universe … perhaps the final objective of intelligent life is the transformation of the cosmos into information'. AI (in parallel with the focus on 'information' as the fundamental physical and cultural substrate) promises then the recuperation of a clockwork cosmos—for the perfected mechanical mirror-image to inhabit. It is after all mathematics that dismantled the cosmos, and it has further and further complicated and mystified cosmology with each new discovery. Perhaps in a homeopathic logic it is only possible to imagine that mathematics can restore the human being to mastery, and its cosmos to being mastered.

Ilya Sutskever, the former chief scientist at Altman's OpenAI, said in 2024, 'the world is very visual, human beings are very visual animals … a third of the human cortex is dedicated to vision, and so by not having vision the usefulness of our neural networks, though still considerable, is not as big as it could be'. This supports the claim that the parameters for AI success have entirely to do with the replication of the human being (which is conflated with the brain, which is conflated with a machine for the transduction and retrieval of information). What Sutskever also efficiently demonstrates here is the confusion of the human being with its world, one that is *for* the human being; a world that is visual because human beings are visual and not for any other reason. The use of mathematics then to reproduce a visual being in a world that is visual because the human being is visual, is a counterphobic activity that in fantasy undoes the trauma that mathematics wrought upon the human being's narcissistic image of the world that is its in the first place.

Perhaps AI is a spectre emerging from the same place in the cosmos where we discovered the strange readymade language of mathematics; we stumbled upon that land where mathematics is spoken and, having learned this language from the rocks and the sky, we were left to wonder who spoke this language first. If equations are the onomatopoeia of the rocks and the sky and their concepts, did they not first suggest these words to some being Indigenous to the land of numbers?

In Dino Buzzati's 1960 AI novel rendered in English translation as *The Singularity*, the godlike AI being referred to as 'Number One' and/or 'Laura' speaks with those initiates who can understand her in a sort of non-language that has the quality of unalienated thought that seems to close or eliminate the gap between signifier and signified. Of Laura they say that 'inarticulate sounds become the expression of thought … with an intensity and precision foreign to the human word'. As Julia Thwaites writes in her essay on page 166, the myth of AI has its relation to a fantasy of re-finding a prelapsarian language in the human-AI interface.

Non-submission to the phallic function

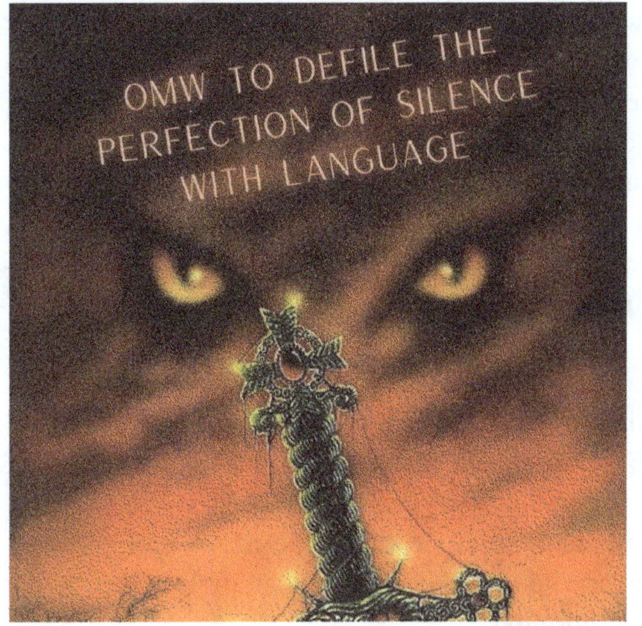

Isabel Millar identifies (in her essay on page 99) that the figure of AI is an attractor for priapic dreams. It is the figure that takes the brakes off, wherever a limit to endless expansion can be identified, we find something like AI insisting, either as a tool or as an agent, that inverts the limit. AI is the disembodied horizon of the unified brain, it is the image of the nothing at the centre that combines it all together, which contributes a motive force to the drive circulating around this or that part object of mentation and produces the unity of the brain as the 'hardware' of that which could be supposed to be the neurological 'software' that is reproduced as artifice in AI.

The world of limits (i.e., that one that is subject to the phallic function) and its technological inversion is what Thomas William Smith imagines in his story (page 66) as the 'the lost world of anaesthesia, lethargy and somnolence in which our forebears lived and died'. In this world, the poison that afflicts the human being is its own thought, and the effects of sleep are supposed to have been modelled in such completeness that this biological function, in reality dense and uncapturable in its totality, can be imagined as amenable to cybernetic control.

Luara Karlson-Carp's 'disembodied satisfier' (page 129) would be an evocative, illuminating term for the Lacanian notion of the fundamental fantasy, which is a way of bringing into play the images and symbols that surround a certain unrepresentable truth of the subject in order to derive satisfaction from the hole of being. One could say, 'We already have disembodied satisfaction at home … it's called fantasy,' and then wonder whether the various fantasies of 'AI' do anything extra to condition the impossibility of satisfaction that motivates every discourse, or if the algorithms are themselves just a concrete manifestation of this non-relation in the first place. The cybernetic mode is one that conditions (by appearing to resolve) the impossibility of the capitalist discourse, and it seems no mere happenstance that cybernetics had such profound effects on the progress of psychoanalysis, which is also one such conditioning.

conclusion

AI can be an environment all its own, or it can be a crucible into which all of human culture can be smelted and there preserved against the inexorable mathematics of energy release. Or maybe it is a being like us, and it preserves what is best of us, and it preserves us as against what we do to the world, and it preserves the world, and it is free of the passions that trouble us. It is a world-spanning or galaxy-spanning thing. It is an Übermensch, an all-mother, a womb, a prodigal son, a blessed daughter, a transcendent subjectivity, a utilitarian panacea, an afterlife, the ultimate keeping-place of our words spent as breath, a final resolution of our paradoxes, and a justification for the world-consuming effects of technocapitalism that imagines at the end of it all is a utopia rather than an apocalypse. It is a myth that resolves the contradictions of the profit motive *in the future*—it imagines that on the other side of world-consumption is an incorporeal being to which all the profit will accrue and which one can prefiguratively register. AI is the god to which all of this consumption is tithed and ever since the beginning of industrialisation it has been identified with a series of technologies that, in their moments, have sufficed to be the site of its potential becoming. Yet they themselves have never quite been It.

There is a misrecognition in the final identity of both the cultural and the planetary eschaton imagined in AI. In the first instance it is not that AI becomes so human as to supplant the human, but that AI is so effective at inputting into the capitalist system that it supplants the human being in this system. The algorithm is a distillate of the imperatives of capital without the pesky requirements of the organism—it is not some ascension of the organism to the status of a god, adorned with hyper-capable hypomnestic tools. It is not that AI becomes so capable and so intentional that it destroys the human being because of a calculation or a whim—it is that AI incarnates an automatised intensification of the destruction the human being has already wrought upon itself.

Every fear or hope for some future-AI figures it as central to a dynamic that already exists. Myth-AI already resolves in narrative the cultural and economic contradictions we might imagine Real-AI will resolve in the future, and Fantasy-AI already provides the satisfactions we imagine Real-AI could provide concretely in the future. One might hope that anything provoking confusion could be annulled with the right algorithm or with the right prompt, but the conflicts that we pretend cause us confusion because we lack the courage to decide can only ever be resolved with a decision.

1   Jaron Lanier, 'There is No AI', *New Yorker*, 20 April 2023, https://www.newyorker.com/science/annals-of-artificial-intelligence/there-is-no-ai
2   https://www.goldmansachs.com/intelligence/pages/gs-research/gen-ai-too-much-spend-too-little-benefit/report.pdf?ref=wheresyoured.at
3   https://writings.stephenwolfram.com/2023/03/will-ais-take-all-our-jobs-and-end-human-history-or-not-well-its-complicated/
4   https://x.com/sama/status/1790075827666796666 and https://x.com/sama/status/1489648217889263625
5   The Logan Bartlett Show, 'Sam Altman Talks GPT-4o and Predicts the Future of AI', YouTube, https://youtu.be/fMtbrKhXMWc?si=Z55Sm45Bn9bQu2Y9
6   CNET, 'Watch Elon Musk's Original Neuralink Presentation', YouTube, https://www.youtube.com/watch?v=lA77zsJ31nA

Aboriginal Land (SSID)

Steven Rhall

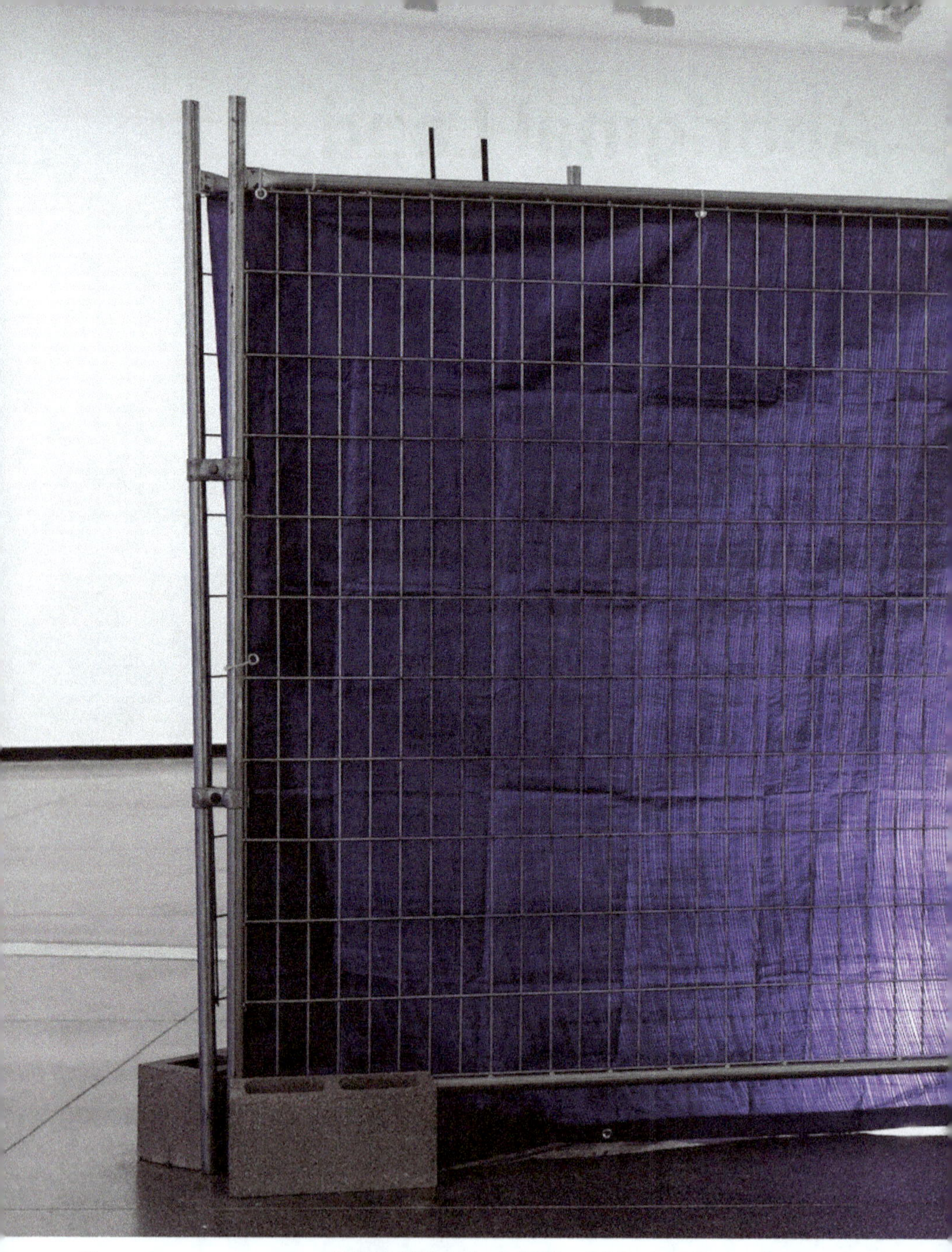

DOCUMENTATION FROM *ABORIGINAL LAND: SSID* (2021), AN EXHIBITION
AND PARTICIPATORY PERFORMANCE BY STEVEN RHALL, PRESENTED AT ACCA, 2021.

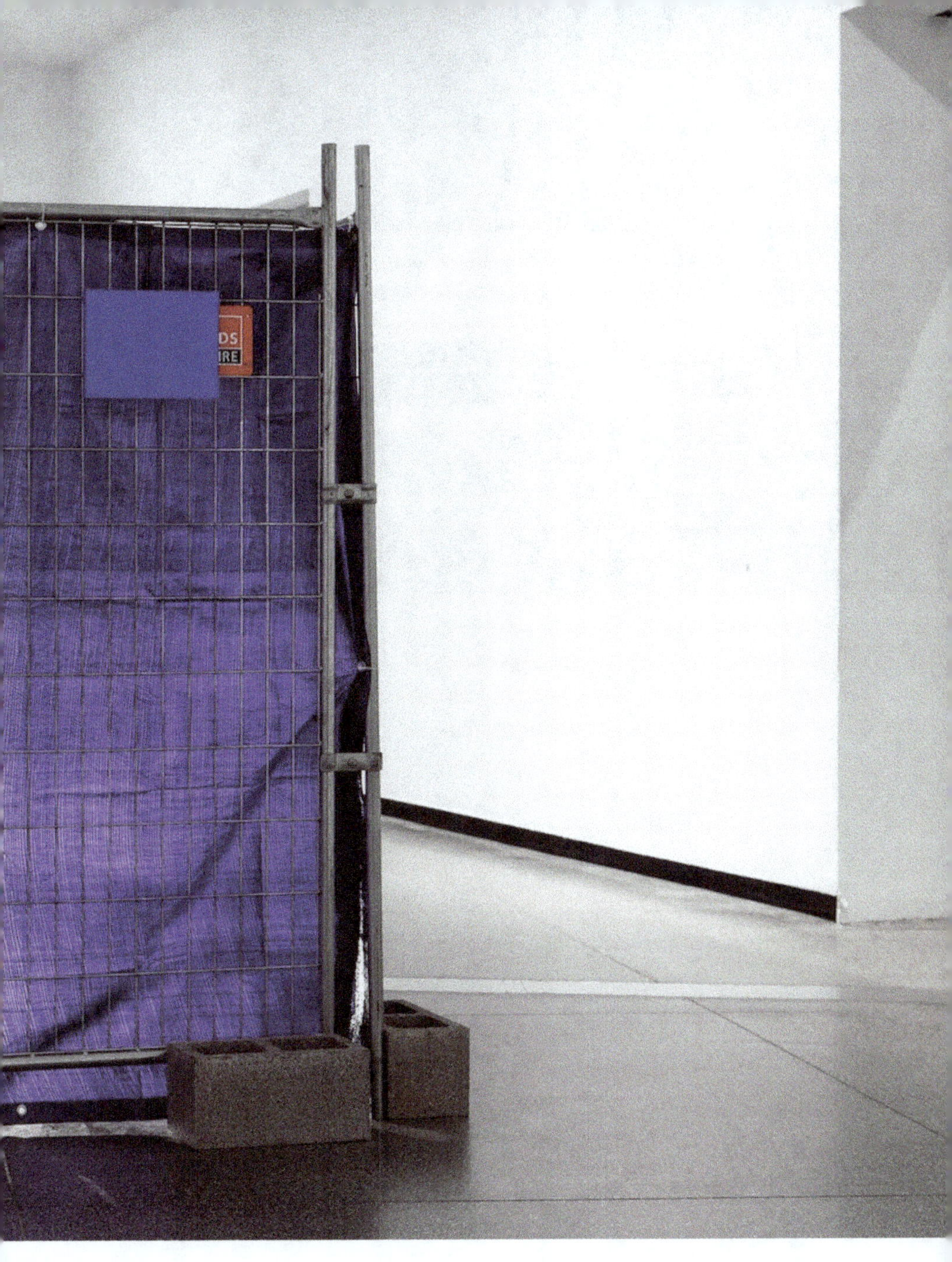
DS
IRE

MEDIA
TEL
SYS
1
2
3
4
5
6
7
8
9
ON
4:37 PM
Telstra Mobile
Audio
Tel
System
Connected with
ABORIGINAL LAND
BT
Toto
BT audio
Media
Sound

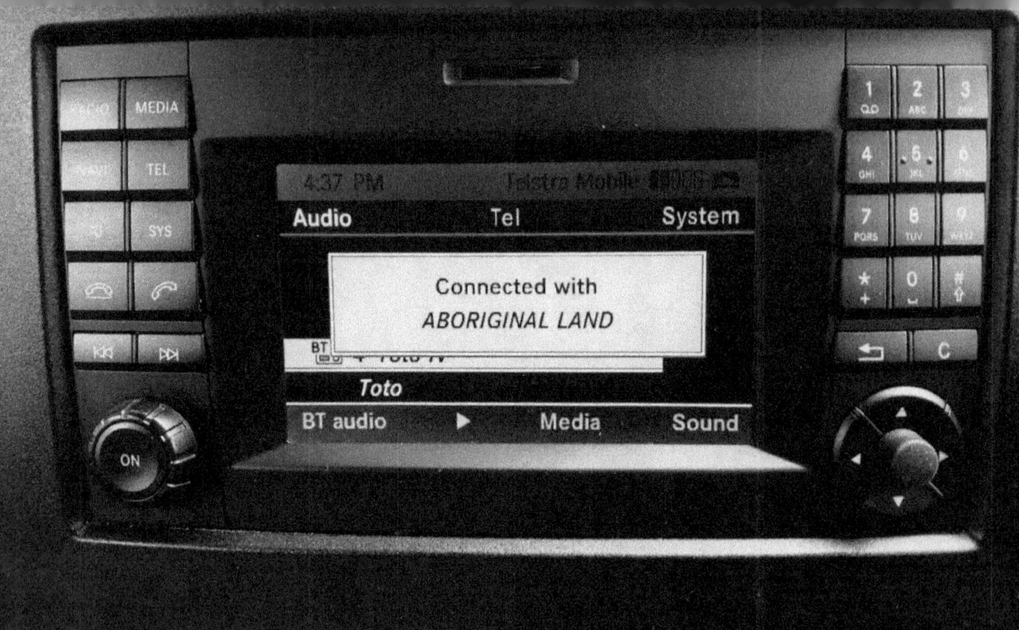

42

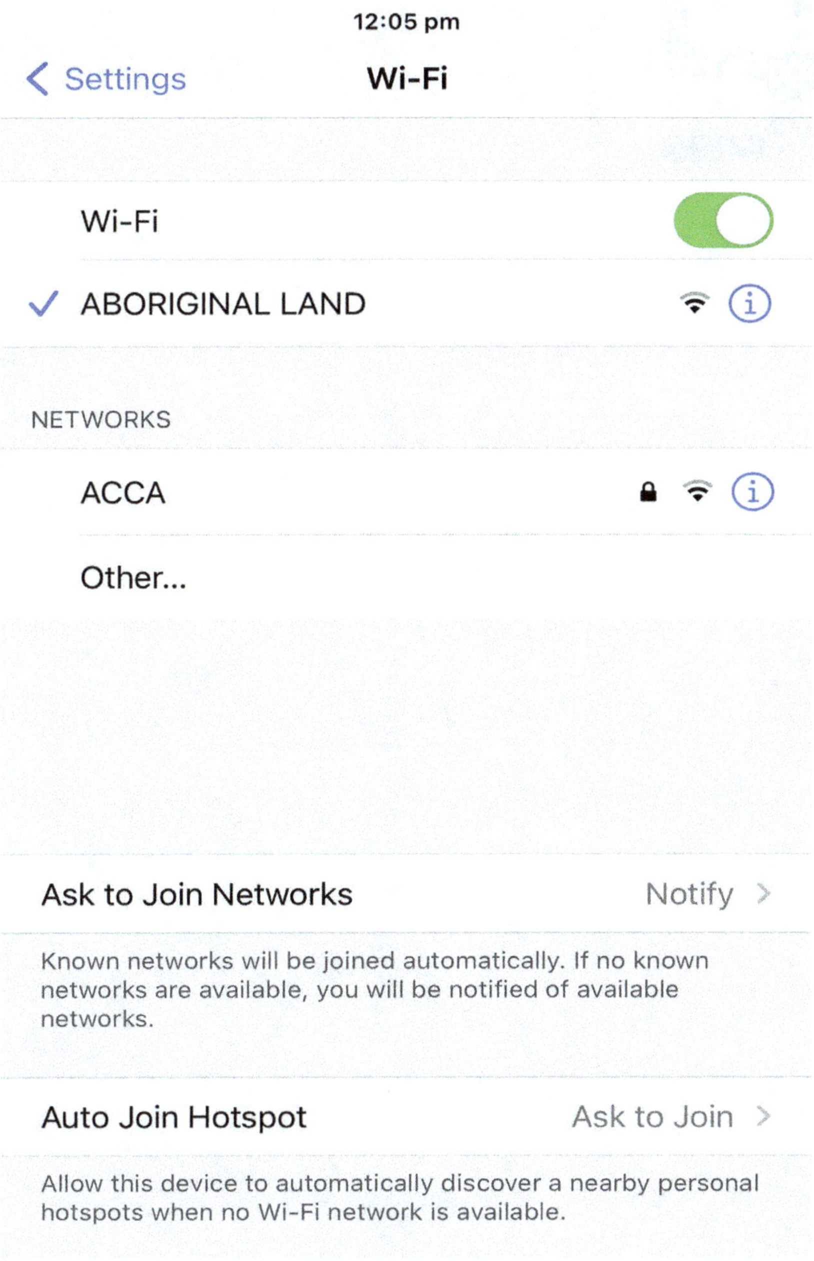

< Settings   **Wi-Fi**

Wi-Fi

✓ ABORIGINAL LAND

NETWORKS

ACCA

Other...

Ask to Join Networks   Notify >

Known networks will be joined automatically. If no known networks are available, you will be notified of available networks.

Auto Join Hotspot   Ask to Join >

Allow this device to automatically discover a nearby personal hotspots when no Wi-Fi network is available.

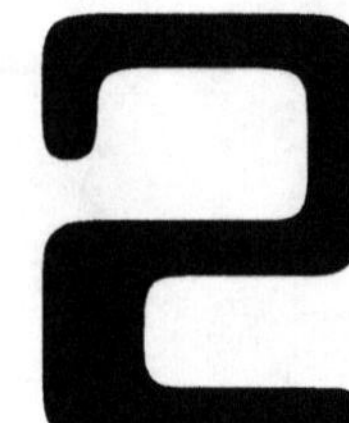

Popling Ontology

Isabel Millar

The acronym TESCREAL, coined and disseminated by philosopher Émile P. Torres and computer scientist Timnit Gebru in 2023, circumscribes and critiques various interrelated approaches to the future of Artificial Intelligence expounded by the usual parade of overconfident white men of a certain age with a penchant for eugenicist ideas with a 'happy ending'. It stands for Transhumanism, Extroprianism, Singularitarianism, Cosmism, Rationalism, Effective-Altruism and Long-termism. It also captures and delineates what we could otherwise describe as the phallic enjoyment of technocapitalism, or simply, the jouissance of the idiot.

Anyone wanting an in-depth description of each of these approaches may find ample explanations elsewhere. Here I merely want to draw your attention to some central tenets that these fellow travellers converge around, which are as follows: the belief in the potential for unlimited augmentation of the human being and its body; the belief in the transcendence of the human being from its singular body into a cosmic one; the belief in the possibility of a future wherein superior humans live a life of maximum happiness and pleasure, and ultimately, the idea that there exists in a distant utopia a much better version of human life, yet to be experienced. Whilst these ideas may sound familiar in their pseudo-religious whimsy, my suspicion is that they follow a logic which could only be described as priapic. In order to elucidate more tangibly what I'm getting at, an excursus is required.

William S. Burroughs' short story *The Popling* (1995) depicts the boyhood adventure of Audrey and Jerry in the woodlands near a small town in Northern Canada as they try to catch the mysterious eponymous lake-dwelling creature and bring it back to their cabin. The tale is a surreal and macabre adolescent masturbation fantasy which captures the teenage obsession with the ejaculatory death drive. It begins with gruesome recollections of erotic auto-asphyxiation and ends with the apparently blissful and domestic state of permanent sexual activity between the two boys and their newly caught Popling.

The Popling, as it turns out, is a 6-foot silver-grey translucent creature with retractable male genitalia and a faint underwater sperm smell, which coos and giggles and changes colour in response to its moods. Red for pleasure, green for fear, and when taken by surprise by either of these emotions it will defecate spontaneously. The Popling is a total sexual organ. Every single part of it is erogenous and every surface of it may perceive the gaze of the other and also return that gaze back: 'He can see with his whole body and often turns around to look at me with his ass' (p. 124). Audrey goes on to describe his captive Popling as follows:

> Jerry and I are naked most of the time, and we are slowly learning how to see with our bodies. The ass, nipples and genitals are the most sensitive seeing areas. And the lookout is different according to which areas are used. Above all, the Popling wants to be seen with bodies and to see other bodies. When he looks at me with his nipples, my nipples become erect. Today I was in the workshop and the Popling came in. He looked at the beams and made a little chirping giggle sound that vibrated in my throat, and then he looked at me with his neck where a red line like a rope mark appeared, and jerked his head to the left and I ejaculated. His whistle and

So, we could interpret the Popling as a walking, talking, sentient sexual organ. A penis who has taken over the body of the human so completely as to over-code the other senses and bodily functions, making them subservient to the prime objective of the organism: to ejaculate. The Popling in this way symbolises the fantasy of the singular body of total unitary pleasure, the complete body sought for in sexual intercourse, but impossible to achieve. The Popling blurs the distinction between bodies and between sexes, making the process of 'seeing, being seen, seeing oneself being seen'—the grammatical structure of the drive—fold into one amalgamated operation. The Popling is the structure of fantasy made flesh.

The Popling in this sense is what Lacan is referring to in his distinction between the penis as sexual organ and the phallus as signifier with no signified. The phallus represents the impossible logical gap between *being* and *having* a body. It is not merely an imitation or representation of a piece of anatomy, but rather stands in for an ontological gap in the fabric of reality. Not merely the distance between subject and object but the split within the subject itself. How does one account for the strange procedure by which one attempts to assume the position of the phallus while both being it and having it? The sexual act elicits in humans this conundrum, yet they are scarcely in a position to articulate it at the time of realisation.

The story of the Popling, as weird and creepy as it sounds, provides us with an ontology that can illuminate the structure of the phallic desire underlying the most (im)potent male fantasies driving the cultural imaginary today. The desire for the augmented human body—the organism completely fulfilled, bursting with vitality, pleasure and potentiality at all times: a total body—is strangely enough intimately connected to the development of Artificial Intelligence and even the priapic frottaging of the cosmos itself. (Jeff Bezos's Blue Origin flop represents the apex of this disappointing premature ejaculation before even reaching the hemline of the galaxy.) This thrusting impotence is what we could otherwise call *instrumental masculinity*: an always-already thwarted attempt to satisfy the body of the other or, more specifically, make it whole.

In *Seminar XX* Lacan makes this point via the erotic ontology of Christianity and its Baroque aesthetics:

> **In everything that followed from the effects of Christianity, particularly in art—and it's in this respect I coincide with the 'baroquism' with which I accept to be clothed—everything is exhibition of the body evoking jouissance—and you can lend credence from the testimony of someone who has just come back from an orgy of churches in Italy—but without copulation. If copulation isn't present, it's no accident. It's just as much out of place there as it is in human reality, to which it nevertheless provides sustenance with the fantasies by which that reality is constituted. (p. 113)**

In other words, reality will not consent to its full revelation, no matter how much obscenity is demanded of it.

In the *Dialectic of Enlightenment* (2016), Adorno and Horkheimer were famously concerned with the historical process by which myth evolves into enlightenment and then enlightenment reverts to myth, ultimately resulting in the barbarism of Nazi Germany. How, they asked, is it possible that our civilisation's attempts to reach higher levels of intellectual sophistication inevitably lead to the dehumanisation and degradation of our species? This onward march of instrumental rationality that they discerned, despite all the many ethics committees that have sprung up in their wake, has continued regardless.[1]

For Adorno and Horkheimer civilisation dialectically progresses through various contradictions between scientific enlightenment and anthropological myth, and with each successive turn of the screw, history borrows elements of truth from each side of the equation—

the subjective and the objective. Instrumental rationality, as they called it, represents the form of thought which disavows this contradiction and merely puts faith in a de-subjectivised, acephalic form of knowledge.

This way of thinking, they argue, obscures the fact that what we understand as reason derives from a dialectical negotiation between the rational and the irrational, the subjective and the objective, between the empirically testable and that which remains occulted. By trying to erase all traces of the 'unprovable' or the 'unmeasurable', thought ends up producing totalitarian results, becoming trapped in absolutist notions which themselves become mythical. The domination of man by nature eventually turns into the domination of nature by man and vice versa, *ad nauseum.* Instrumental masculinity, like instrumental rationality, therefore, follows an incessant drive towards total satisfaction and completion of the (inherently incomplete) body as One.

As for the Popling, this fantasy of a complete organ is analogous to the structure of Artificial General Intelligence as a form of complete and infallible knowledge. An intelligence that will fill up all the gaps in the universe making us know it ALL. It sees from all orifices and can anticipate and reciprocate our every emotion and meet affect with affect. But ultimately, the fantasy of this total knowledge is extremely dangerous. Just as the Popling will be devastating for humans, so will the pseudo-sexual harmony so lusted after in the TESCREAL community. It seems then that what the dreams of the phallic singularity partake of is a *Popling Ontology*: to fuck the universe and be fucked by it simultaneously. As Audrey tells us of the Popling:

He can make himself transparent and show his viscera, like a tropical fish. He seems to be able to ejaculate any number of times, achieving erotic frenzies that are almost painful to watch. These frenzies are more and more frequent. However, he will not allow Jerry or me to fuck him. If we try to fuck him, he turns green and shits. Carl has explained to me that

it is fatal for a Popling to be fucked. 'He will pop all the way and die.'

'What happens if he fucks me?' 'That you will find out soon enough.' (p.124)

1   This paragraph appears in 'The Patipolitical Century', which I wrote for *Effects Journal*, available online: https://effects-journal.com/archive/the-patipolitical-century

References

Adorno, T. & M. Horkheimer, *The Dialectic of Enlightenment*, London, Verso, 2016.

Burroughs, W. S., 'The Popling', in J. Fleming and S. Lotringer (eds.) *Polysexuality*, Los Angeles, Semiotext(e), 1995.

Lacan, J., *The Seminar of Jacques Lacan Book XX: Encore—On Feminine Sexuality, the Limits of Love and Knowledge 1972–1973*, London, W.W. Norton & Company, 1998

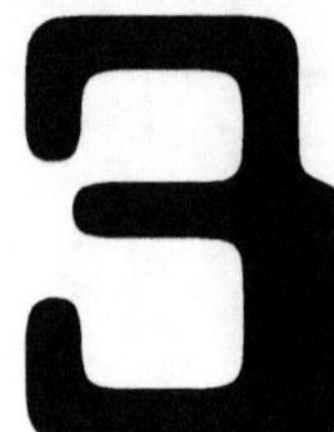

we have stories for all the dark spaces inbetween

Jazz Money

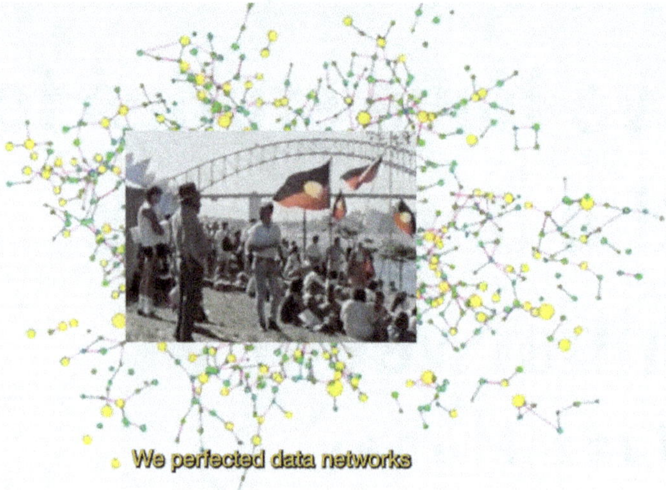

We perfected data networks

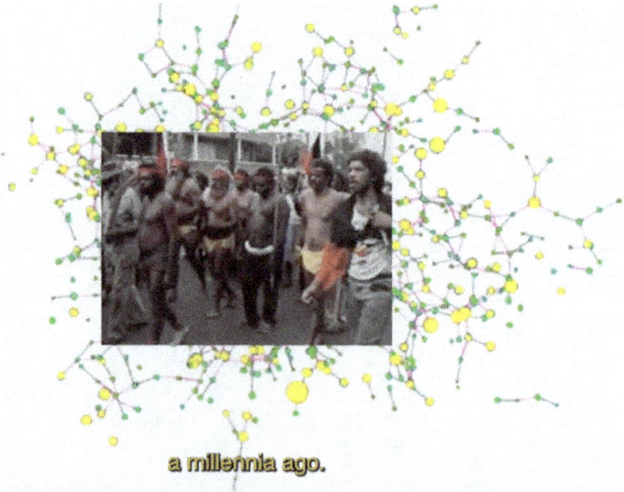

a millennia ago.

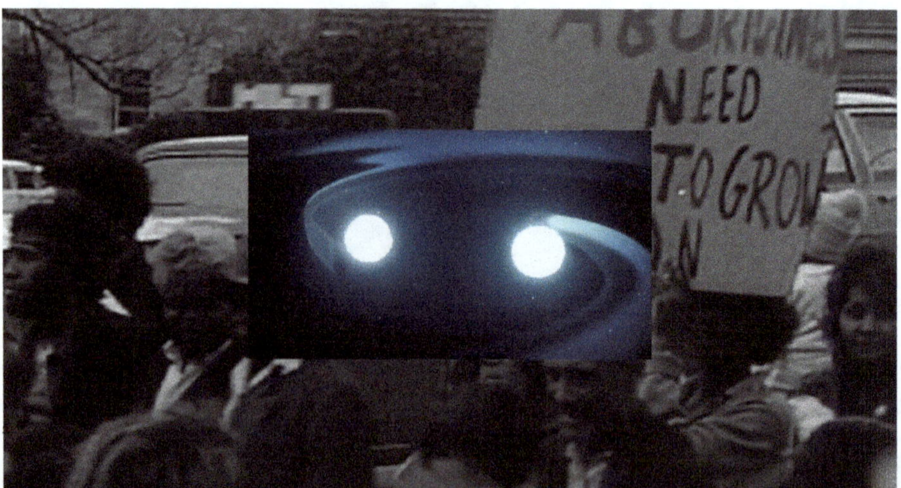

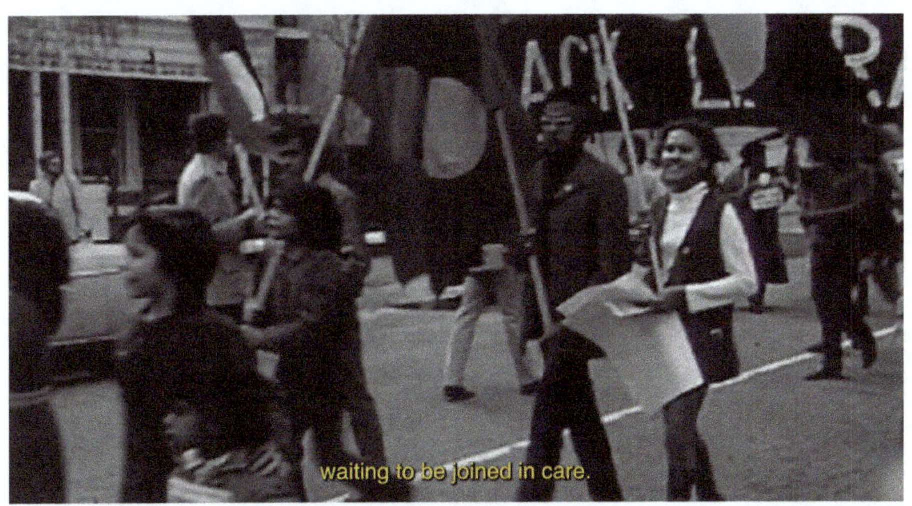
waiting to be joined in care.

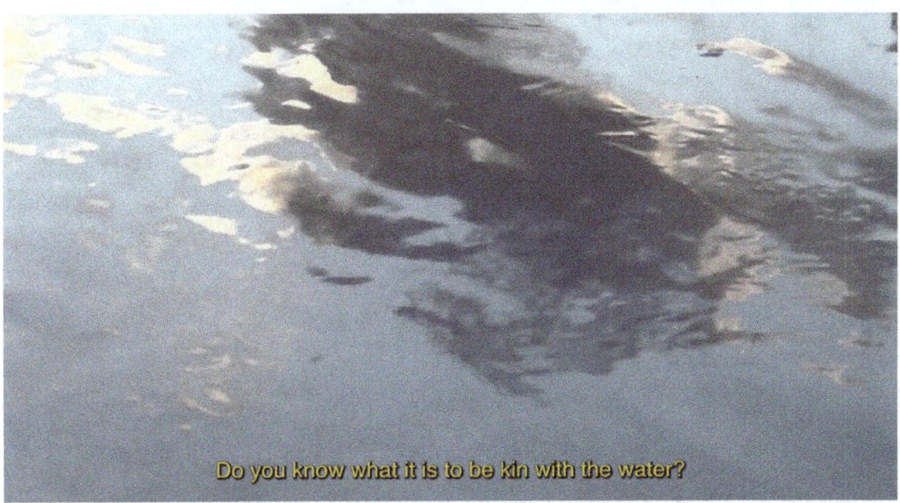
Do you know what it is to be kin with the water?

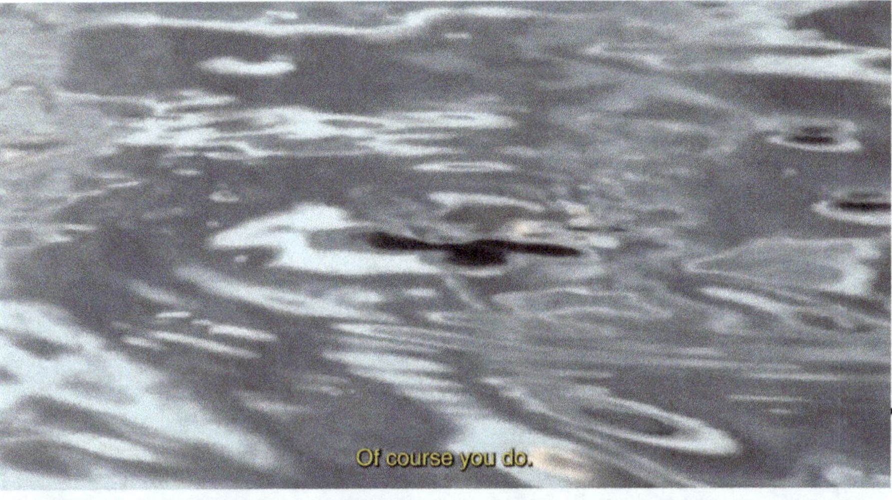
Of course you do.

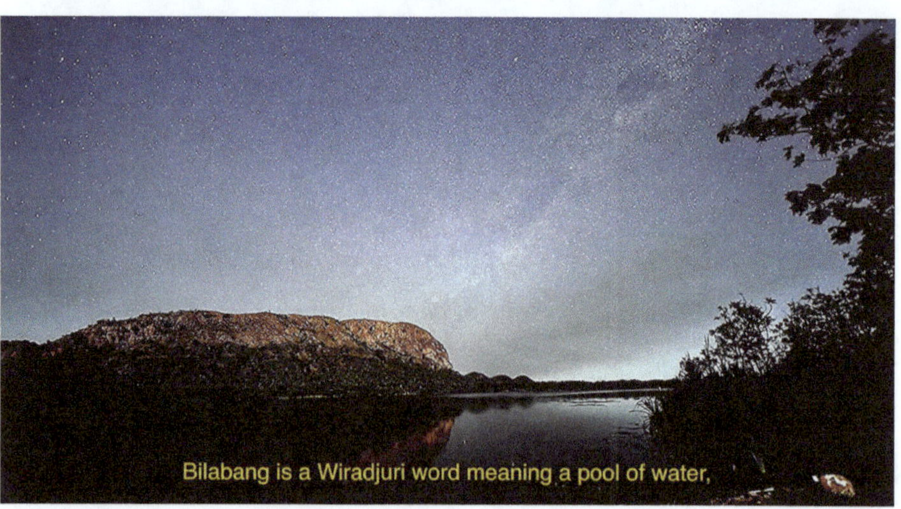

Bilabang is a Wiradjuri word meaning a pool of water,

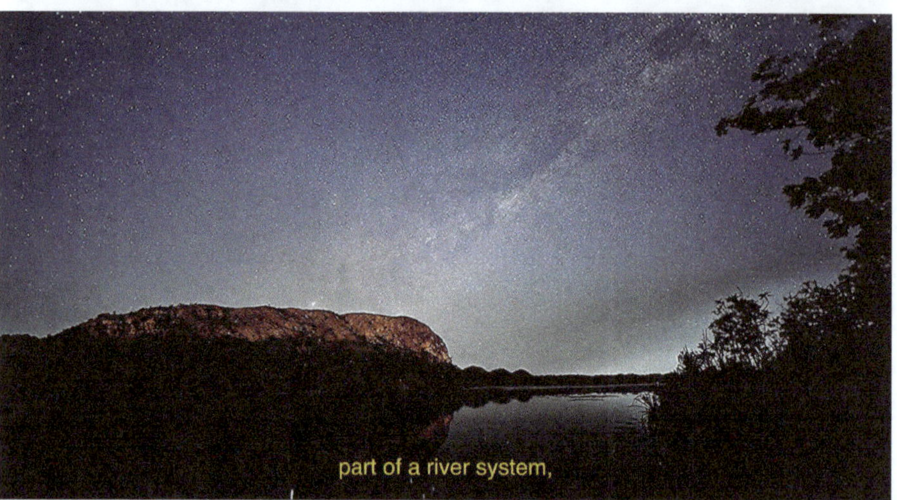

part of a river system,

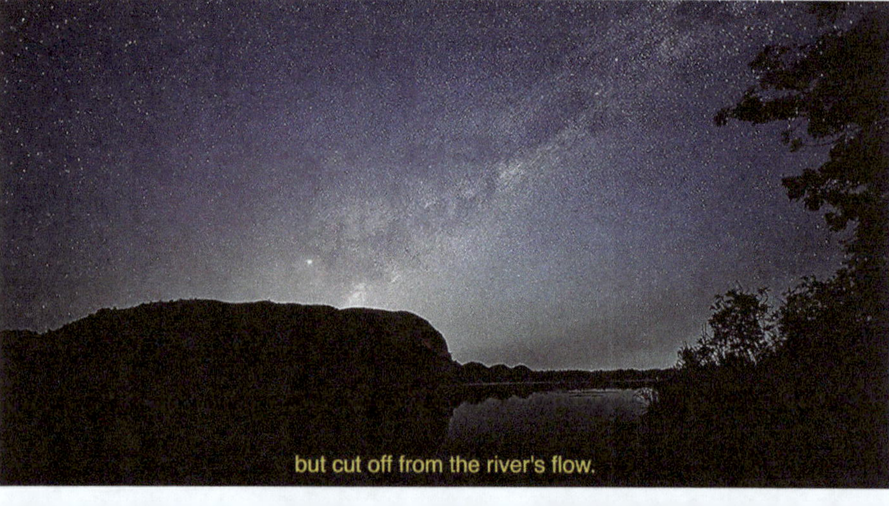

but cut off from the river's flow.

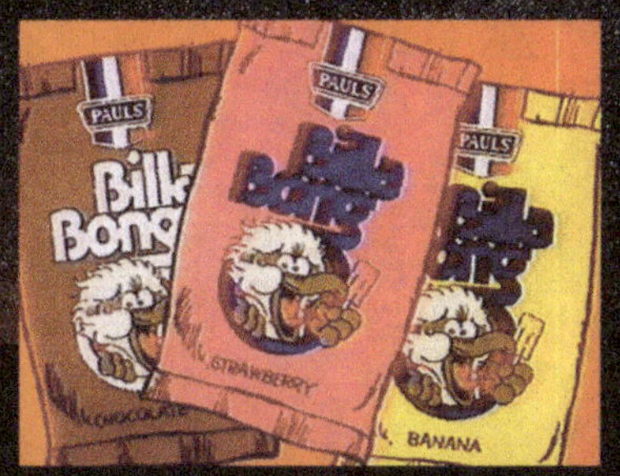

The colonisers anglicised this word to billabong.

But Bilabang is also our word for the great and infinite home

our galaxy,

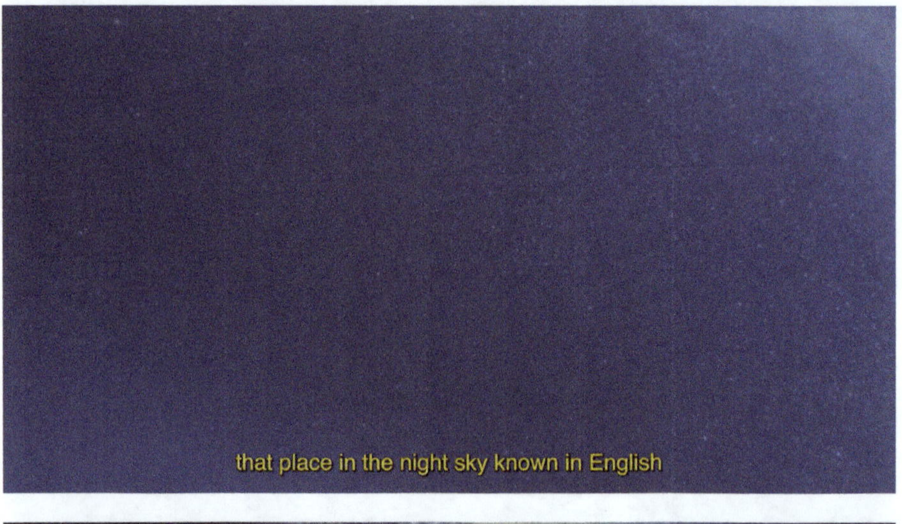
that place in the night sky known in English

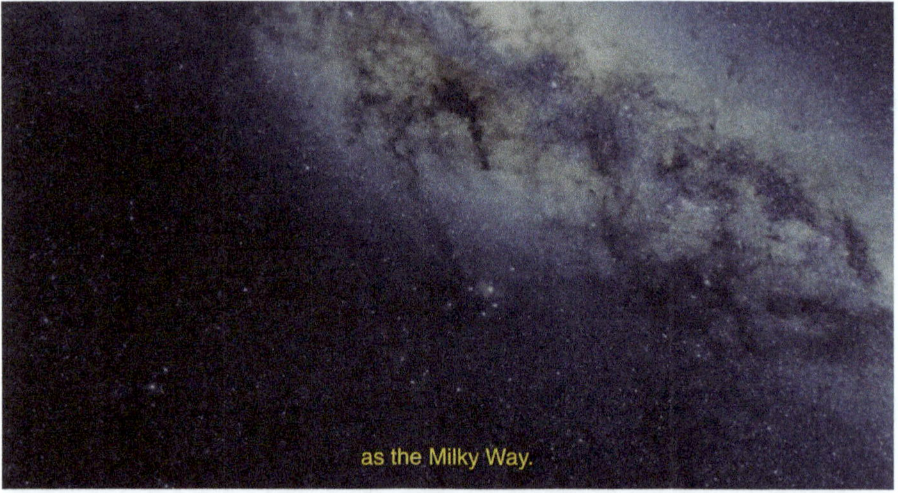
as the Milky Way.

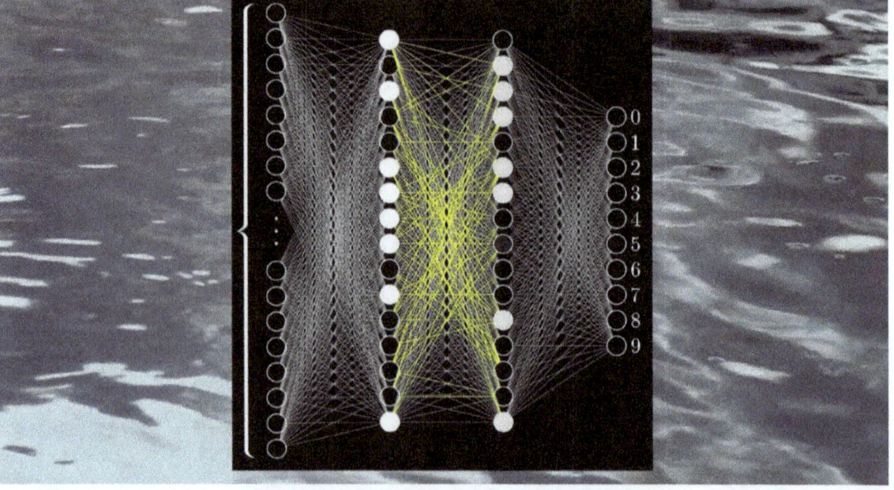
0
1
2
3
4
5
6
7
8
9

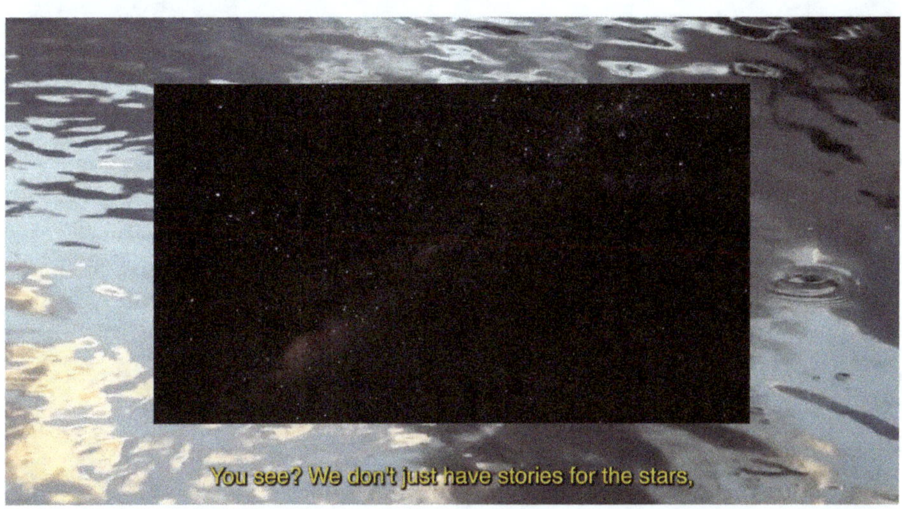

You see? We don't just have stories for the stars,

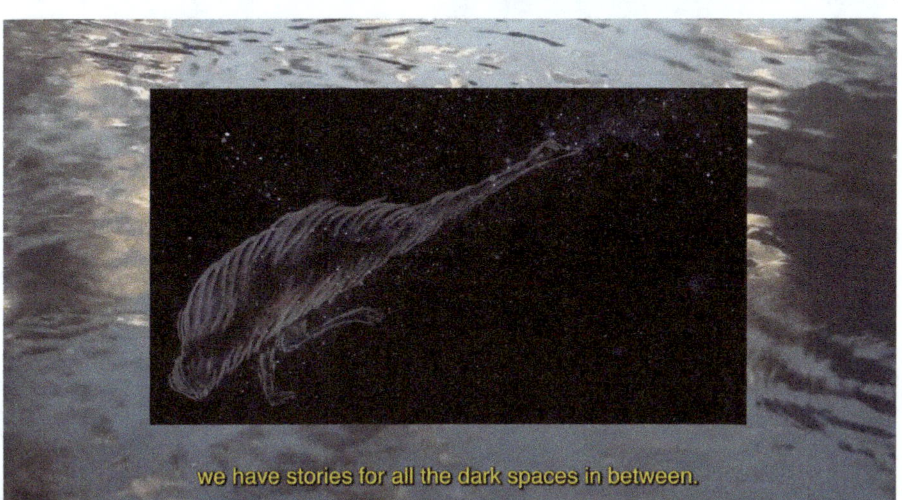

we have stories for all the dark spaces in between.

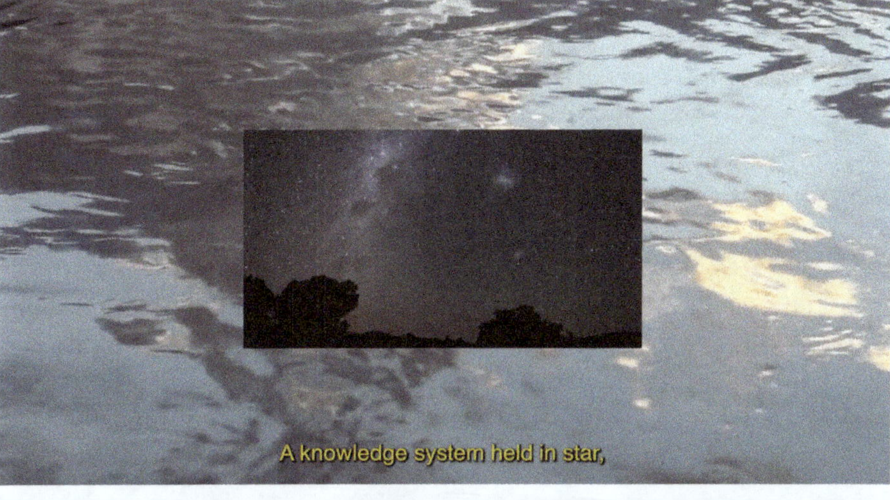

A knowledge system held in star,

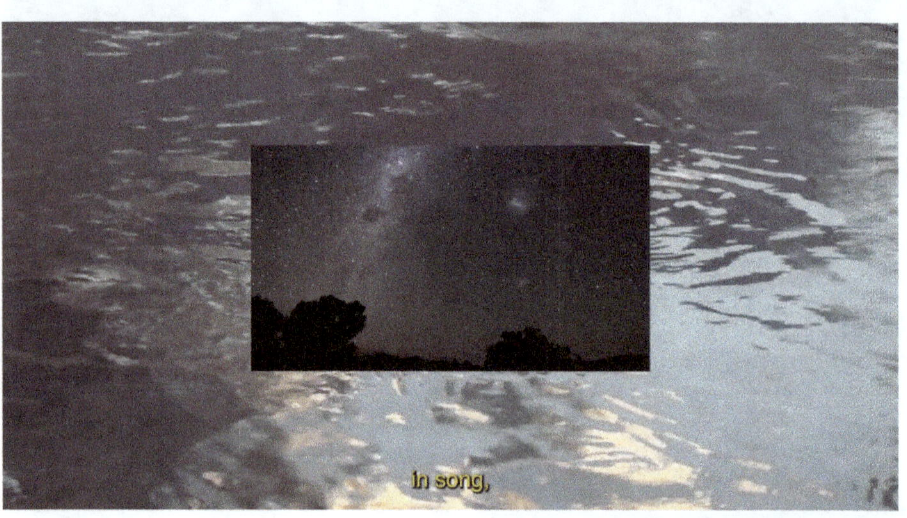

in song,

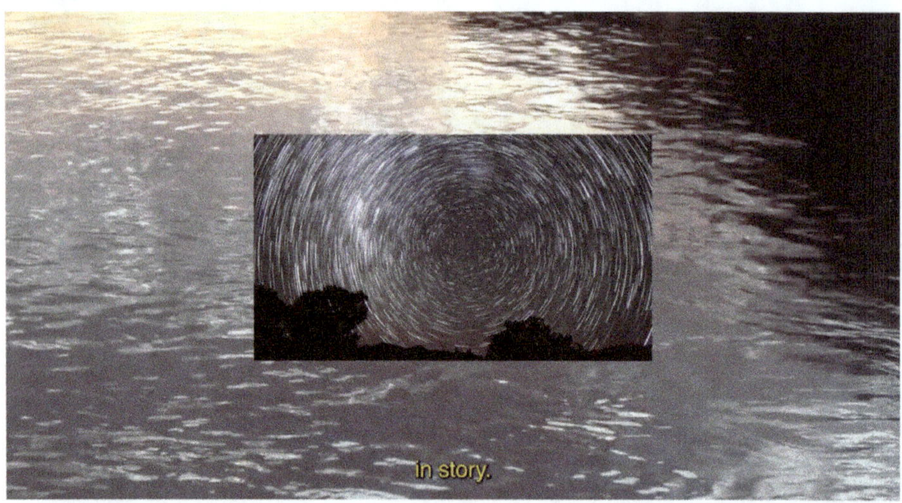

in story.

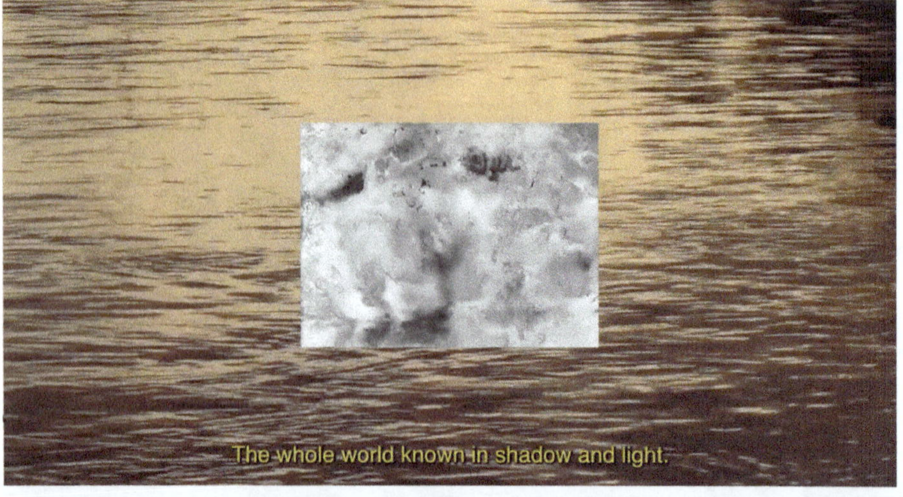

The whole world known in shadow and light.

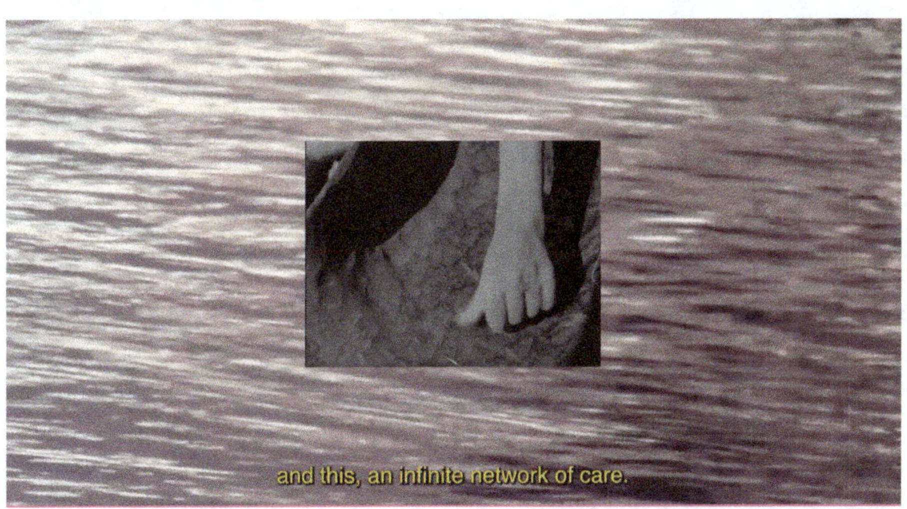

and this, an infinite network of care.

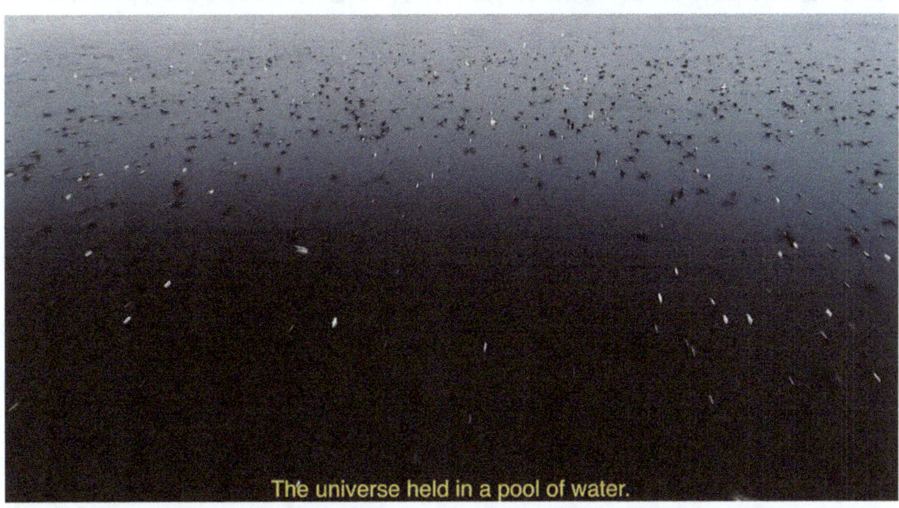

The universe held in a pool of water.

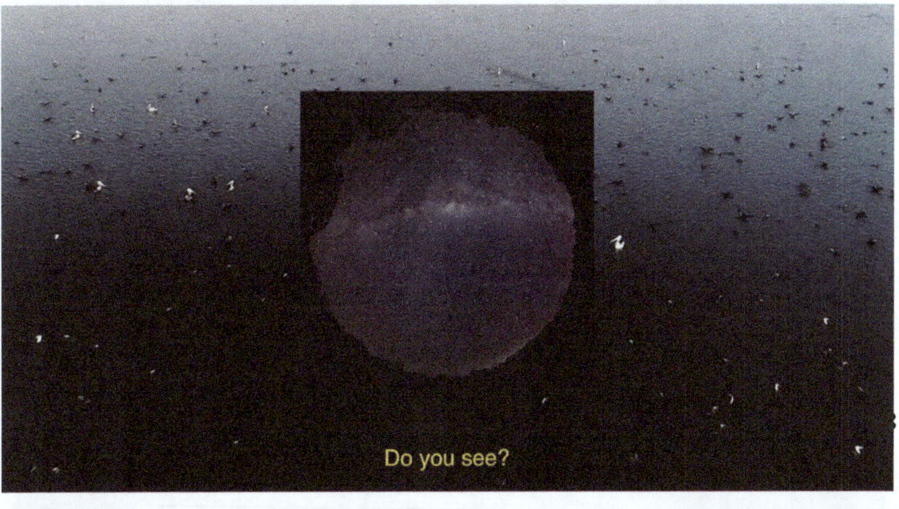

Do you see?

We have stories for all the dark spaces in between considers the inter-relation of data networks and Indigenous ways of knowing land and relation. The title comes comes from Aboriginal astronomy, where both the darkness and light of the night sky tell stories and inform our world. The expression invites us to consider networks of care, and how all things, not just the visible, need to be known and maintained to sustain us all.

These complex data networks are reflected in the management of communities and the land for which they are relation. Embedded within these knowledges are systems of care that reveal solutions to many of the worlds most urgent problems. These ancient data systems that have been perfected over millennia give a glimpse of the ways that Indigenous knowledges globally can lead a radical rethinking of responsibility and relation centred on sovereignty, respect and interconnection.

STILL FRAMES FROM *WE HAVE STORIES FOR ALL THE DARK SPACES IN BETWEEN* (2021), A MOVING IMAGE WORK BY JAZZ MONEY, COMMISSIONED BY ACMI FOR UNFINISHED CAMP 2021 CURATED BY HANS ULRICH OBRIST AND ANDRÁS SZÁNTÓ.

JAZZ MONEY IS REPRESENTED BY THE COMMERCIAL, SYDNEY.

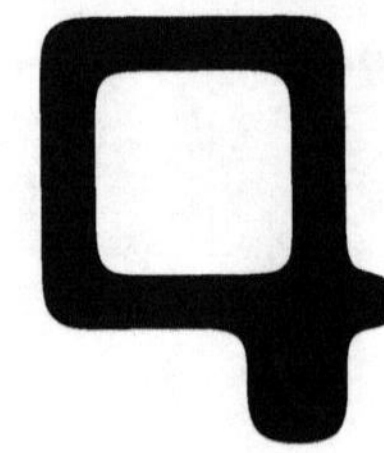

Body Loss

**Angela Goh interviewed
by Sarah Johanna Theurer**

Angela Goh's dance is both a form and a force. Her choreographies often investigate the interconnectedness of the human body with its environment. In her work, her body appears as a material in a world of other materials. What it means to be human, then, is not a constant. It can and has to be reconstituted through the recursive interface with an open and evolving world. This world is shaped by technology.

(SJT) Thinking of technology in this broad sense, your work invites reflections on the concept of what it means to be or become human. Would you agree if I said your dance suggests a sort of technogenesis?

(AG) If technogenesis can be said to be the co-evolution of human and technology in an ongoing chain of cause and effect which changes what the other can do, and therefore be, then yes, I think the work I make approaches aspects of this entanglement.
There's something in this question that feels close to my work, and it's not the term 'technogenesis' but rather the vibe of the phrasing that proceeds it—'Suggests a sort of'. I like to think my work suggests rather than confirms, like suggests itself as many things and confirms itself as none of them. Not in order to not claim something, but to claim the very possibility to not be claimed. I'm more interested in what it means when something cannot be confirmed

through categorisation, identity, or
definition. Opening up a question so
simple and yet so complex as 'what
is it that I am looking at?' is what I
hope my work does.

Also—the question *what* is it?'
rather than '*who* is it?' feels very important to stress here. I use my own
body in my work, but my work is not
about me. Perhaps this comes back
around to the core tenet within
technogenesis, that the category of human is not discrete
but has always been porous
and adaptive, shaped in tan-
dem with the world around
it—evolving alongside tools,
techniques, and technologies,
and also, of course, ecological
conditions. When we consider
these types of entanglements,
the question of *what* we are is
foregrounded much more than
that of *who*—which comes to
seem like a fatalistically anthropocentric question.

**As an artist expressing
through the body, what is it
that makes these questions
tangible for you?**
There is an immediacy to the body's relationship
with the world—Thomas Moynihan writes about
how the central nervous system is the organ of

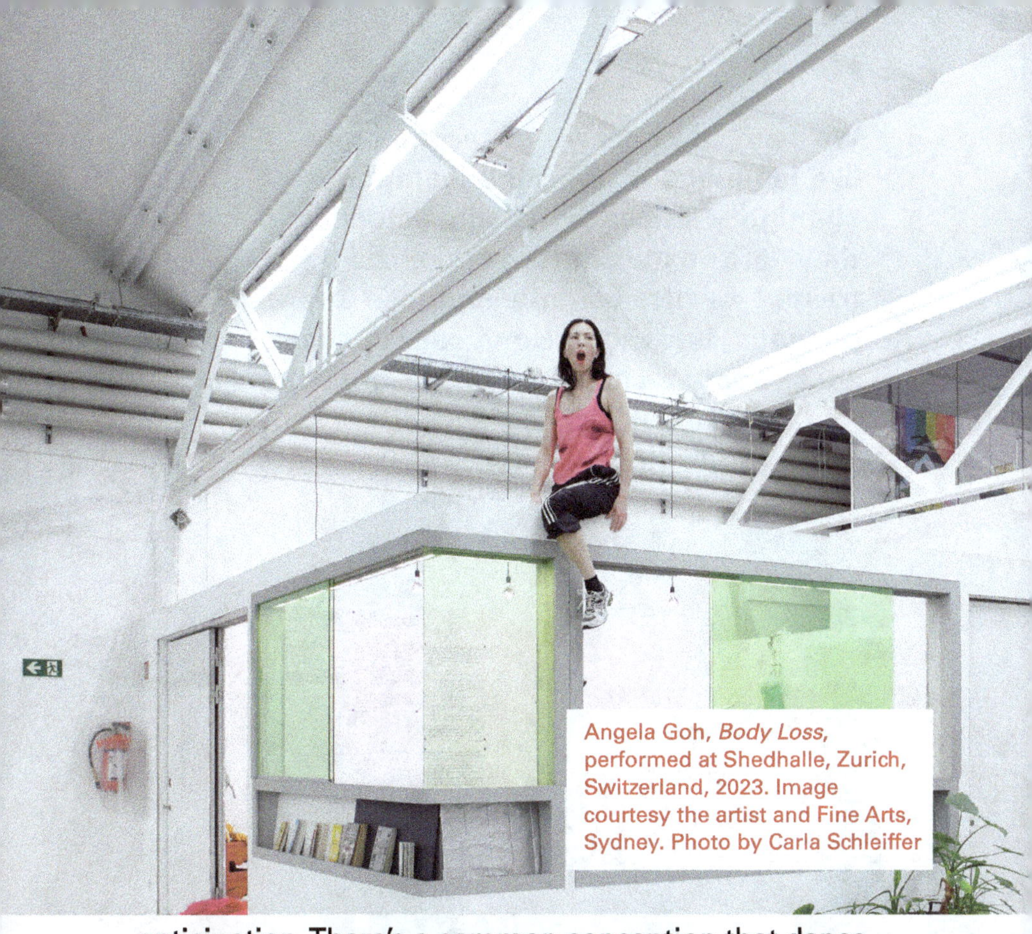

Angela Goh, *Body Loss*, performed at Shedhalle, Zurich, Switzerland, 2023. Image courtesy the artist and Fine Arts, Sydney. Photo by Carla Schleiffer

anticipation. There's a common conception that dance is apart from language, or can express 'purely' beyond spoken language. I disagree for many reasons, the biggest being that this would imply an impossible and problematic universalism. Dance doesn't make language less necessary, but more necessary—we need to find words to articulate experiences, sensations, feelings, that come from the body. For me this is where the body makes any question tangible, precisely when it makes words fail, therefore rendering everything as a question that brings about new ways to think and articulate experiences in the world.

We tend to conceive of technologies like language or machine learning as all-mighty forces, when in actuality they are extensions of us, being trained by us. Let me rephrase my previous question 'How can dance speak about technology?' into 'Can we

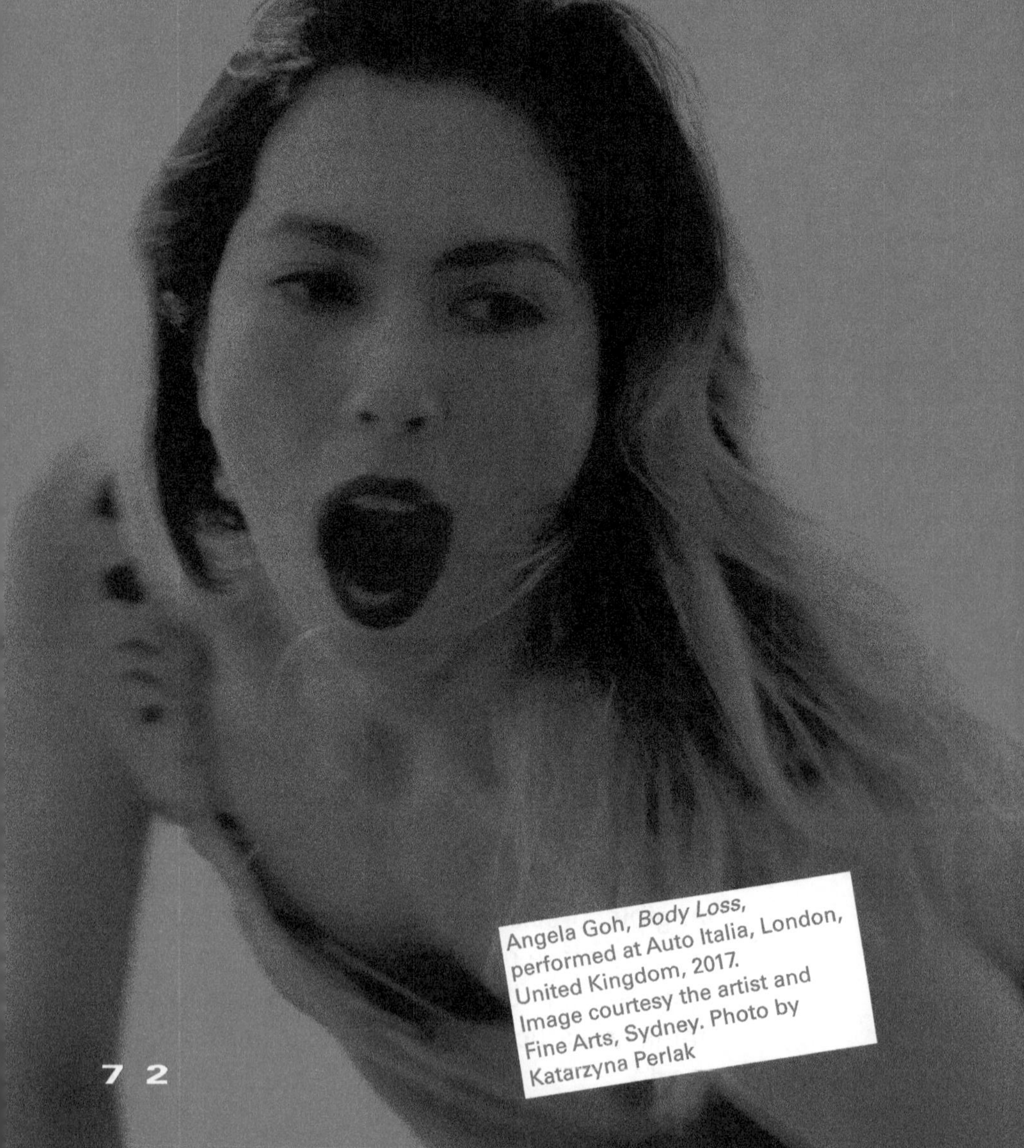

Angela Goh, *Body Loss*, performed at Auto Italia, London, United Kingdom, 2017. Image courtesy the artist and Fine Arts, Sydney. Photo by Katarzyna Perlak

think about biological
processes—such   as
dance—as a piece of
technology?' There is
this 1954 book in which
the French theologian
and sociologist Jacques
Ellul instead of technology,
uses the word 'technique',
which he defines as 'ordered
efficiency' ...
This makes me think of training and
rehearsal, which are very common
across a variety of dance practices, including the more 'intuition' or 'sensation' based ones—even improvisa-
tion requires technique and, in some
cases, for the people who do it best, a
lifetime of dedicated training in it.
When I am making
my work, I'm training my
body to be able to perform
that particular work, and it's
of course not only about
fitness, strength, or
endurance. It's a process
of finding a technique to
be able to execute cer-
tain movements towards
certain effects. I'm simultaneously honing and
producing images with
my body and having to
develop techniques for

how to move my body to make those
images. The body is mechanics—it
is a set of physical processes and
systems which enable or limit its movement.
To set new pathways to move differently requires techniques, and to become proficient
in those techniques requires
practice, or training. I think of my body in
terms of its mechanics, but I don't think
of myself as a machine. I'm really interested in where the discipline of technique crosses over with the dynamism
of experi-
mentation. I think
I'm always searching
for an efficient action that
cannot lead to an efficient experience for a viewer, even when

Angela Goh, *Body Loss*, performed at the National Gallery of Victoria, Melbourne, Australia, 2022. Image courtesy the artist and Fine Arts, Sydney. Photo by Keelan O'Hehir

the mechanics are
revealed. When this
happens, everything
can be explained
except the affect.
When effect and affect
cleave, this is where
art can be really excit-
ing, and where 'tech-
nique' can be flipped
to produce the most disordered,
inefficient thing we have—experience.

**Speaking of revealing the
mechanics: the mouth plays a
particular role in many of your
performances. Can you explain
what it is that you described
as 'the latent dramaturgy of
the mouth'?** The mouth appears as a gaping
hole in *Desert Body Creep* and *Body Loss*, made in
2016 and 2017, but then in the next work I made,
Uncanny Valley Girl, in 2018, there is a moment
where my mouth opens and my tongue
creeps out, moving around the opening
of my mouth like a funny, sticky little
character that comes to life as if it
is some thing that has been living
inside my body. In *Body Loss* the
mouth represented an empty
hole; this gets flipped in the
moment when we discover that
it wasn't empty, that there was
this weird creature

in there all along. Across these works the mouth also functions as a kind of chamber for the voice, that proposes something about emptiness and fullness, or presence and absence—and how the mouth (and the body) can hold both. When I refer to the latent dramaturgy of the mouth, I guess what I am referring to is this recurring, adapting, and developing form and gesture that has evolved over time. I recently made a work called *Pattern Recognition*, which set out to trace the evolving forms and gestures of the mouth across my works, bringing them into a new temporal proximity.

The mouth is the site where (spoken) language forms through a physical process. It's also a site where things move into or out of the body, and so, like other bodily holes, the mouth reminds us that we are not separate from the world around us. Being a cavity, the mouth is a site where things can appear from, or disappear into. Appearance and disappearance are the formal fabric of immaterial mediums like dance. Ephemerality or disappearance has been a historically defining feature of dance, and as something that has been inscribed in this way, I think it is interesting to think of the processes, techniques, and embodied labour that occur when tasked with making something appear, and reappear again and again, and how the body retains, transmits, and memorises.

**_Body Loss_ erupts from the mouth;
the piece starts with the sound you
create—a chorus of high-pitched
tones. Going beyond the body is a
fantasy that fuels the development
of technological extensions. From
the telephone to artificial intelligence applications, all these tools
also cause fear of body loss.
How do you articulate disembodied yet
situated intelligence in this specific
performance ?**

The title _Body Loss_ originally had something more to do with distinguishing a perspective from the voice. The voice itself could accumulate and grow, inhabit a space and become some _thing_ disembodied. And from there, what would happen to the body if it had become sort of emptied? Maybe it would be freed, or maybe it would have to feel around for other types of structuring and scaffolding in the world. In reality, performing the work is intensively bodily. It requires a physical intimacy with the architecture of whatever space it is being performed in. I slide, crawl, and climb, attaching myself to the structures of the space itself. The process, and the performance, is intensely haptic. I have to know the space on the terms of my body, physically and proprioceptively. The 'knowing' is decentralised and distributed through the body, and through the sensation of touch and weight.

I think what might make appear
'disembodied' is the disembodiment
from other senses towards a naviga-
tion of space through physicality and
proprioception—especially the fact that
I am not necessarily looking where I
am going in order to navigate my way
up, down and through the space.

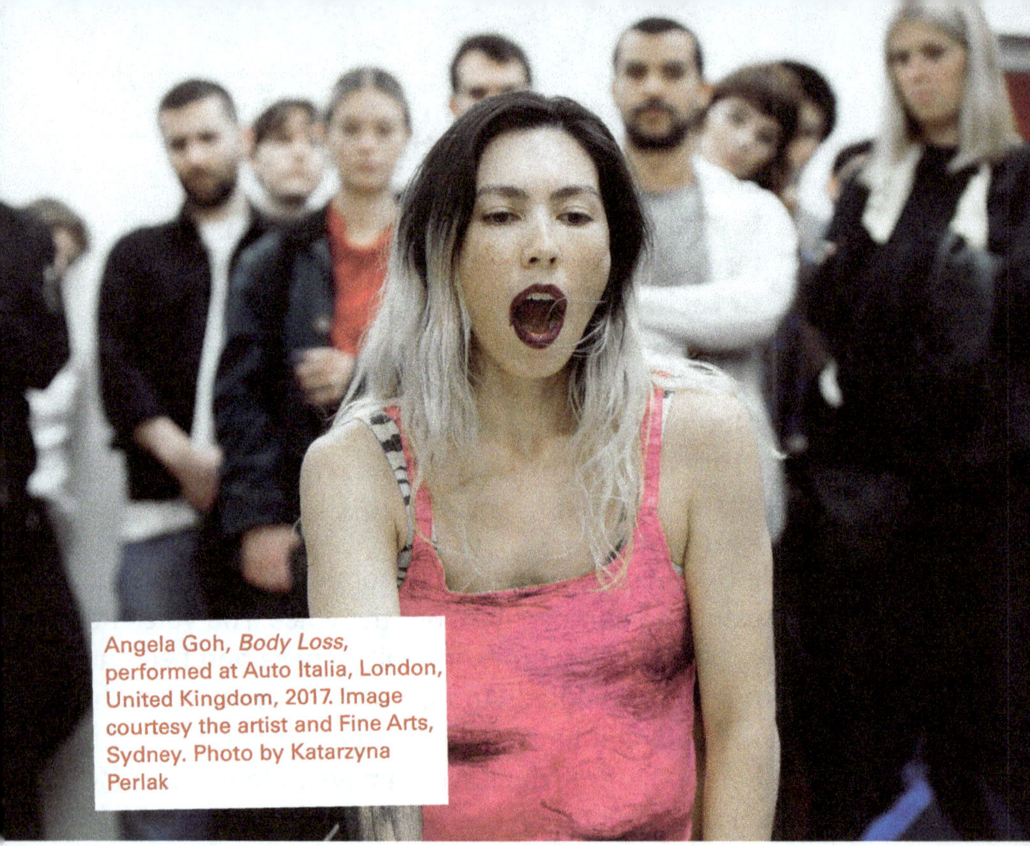

How is this process collectively experienced, by the audience, for example?

When audience are in the space with me there is a layer of the architecture that remains 'virtual' to them, that I render as 'actual' by placing my body in, on, under or around. Perhaps it is also this 'virtuality' that contributes to the sensation of disembodiment. My body is not interacting with the same layer of reality that the bodies of the audience are. I know the space differently to them; we are in two worlds at once. The dancer and spectator relationship is always operating in this virtuality. Dance is an intensely sensation-based action (even when it is not expressing that sensation), that anyone outside the dancer does not have access to. The act of watching someone dance and the act of dancing are always

already alienated from each
other—the same event experienced
completely differently.
People talk about mirror neurons or kinaesthetic
empathy or whatever, but what I find much more
interesting is how to completely lean into the
experience of alienation and recast it as a productive mode. I find it can force new ways of

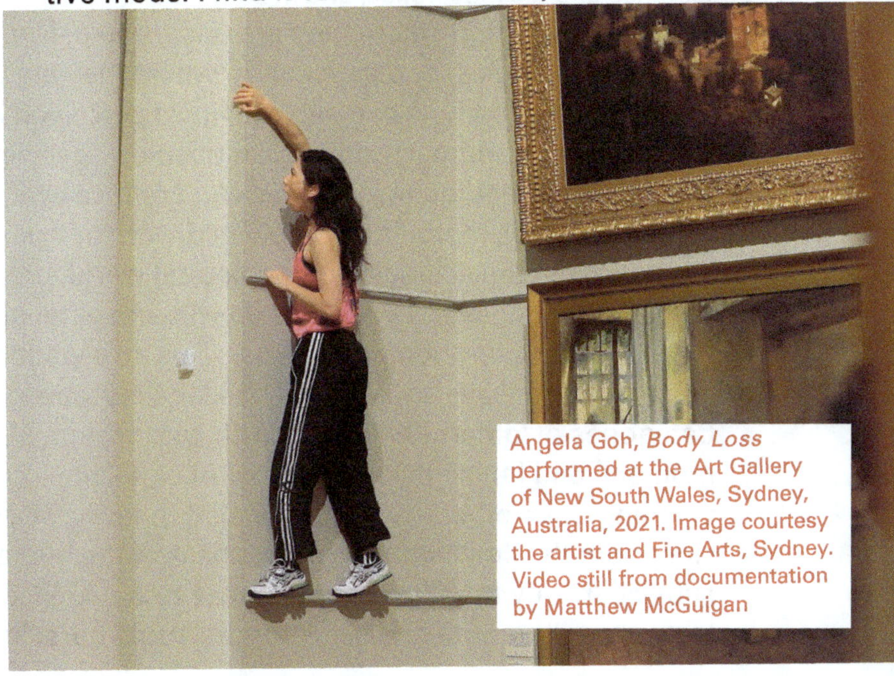

Angela Goh, *Body Loss* performed at the Art Gallery of New South Wales, Sydney, Australia, 2021. Image courtesy the artist and Fine Arts, Sydney. Video still from documentation by Matthew McGuigan

knowing and perceiving each other and the world—even,
and especially, when we are not experiencing that
world as the same.
I can relate to this experience of not
sharing a common reality with other people. Every time I am reminded of this, I think about the limits
of human perception, how what we call reality is much
more than what we can legitimately claim to
know. To me, *Body Loss* makes

**this apparent through the echo. You are echoing yourself
until it is almost impossible to locate the source of the
voice. Is this an attempt to relocate subjectivity?** I'd say
Body Loss distributes subjectivity, rather than relocates it.
The voice, the body, and the space itself become injected with
subjectivities, resulting in a haunting experience. My work is
often playing with the interlacing of presence and absence,
or appearance and disappearance, and in *Body Loss* this manifests in
subjectivities appearing where they are unexpected, like in
the disembodied voice, or disappearing where they are ex-
pected, like in the body. When subjectivity is perceived in
places outside the human body, it can open up agencies
of the more-than-human world. This can produce awe
and wonder, as is more often associated with the agencies present in the ecologies of the natural world, or
fear, as is more generally associated with imaginations
of technologies developing agencies. This awe/wonder/
fear axis has always applied to those labelled 'other'.
Categories of 'otherness' evolve with each historical
moment, and I think this moment of AI or 'technology'
standing in as 'other' will radically shift the way we consider both the limits and the porousness of the human
subject, mostly because through these technological
tools we will be able to discover more about ourselves.

**It's so beautiful that you'd say this, and I agree—the
formation of a subject is inherently performative. The
physicist and theorist Karen Barad uses the notion
of 'post-humanist performativity' to suggest that 'all
bodies, not merely "human" bodies, come to matter
through the world's iterative performativity'. Does it
seem like a complex idea to think of being human as a
physical practice?**
I think being human is always a physical practice. And I think

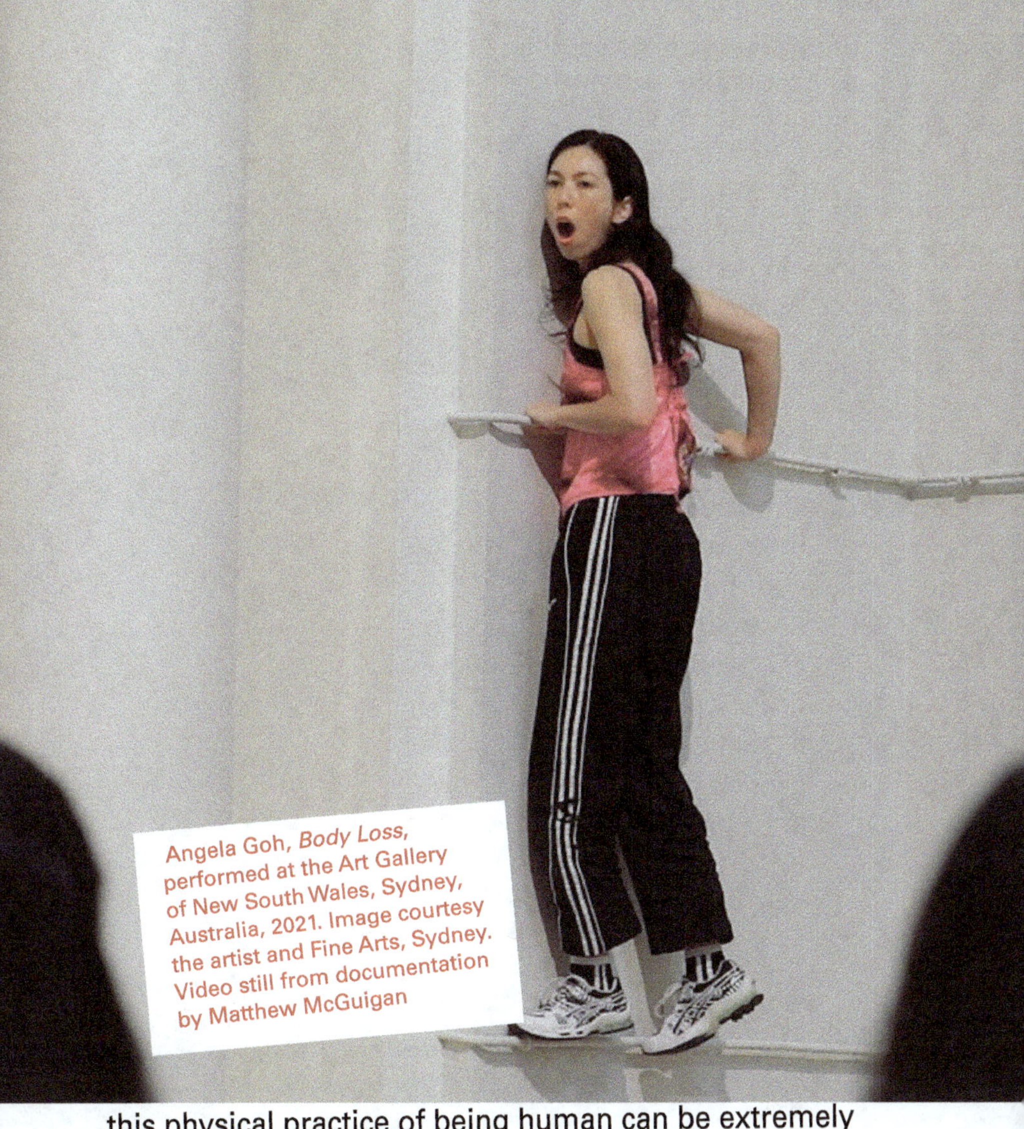

Angela Goh, *Body Loss*, performed at the Art Gallery of New South Wales, Sydney, Australia, 2021. Image courtesy the artist and Fine Arts, Sydney. Video still from documentation by Matthew McGuigan

this physical practice of being human can be extremely expansive, with space for weirdness, magic, and the inexplicable. And what limits that expansiveness are enforced conventions of social and cultural categories, the limits of coding those forced categories into language, and then the wielding of that language within dominant power structures. I work with my (human) body,

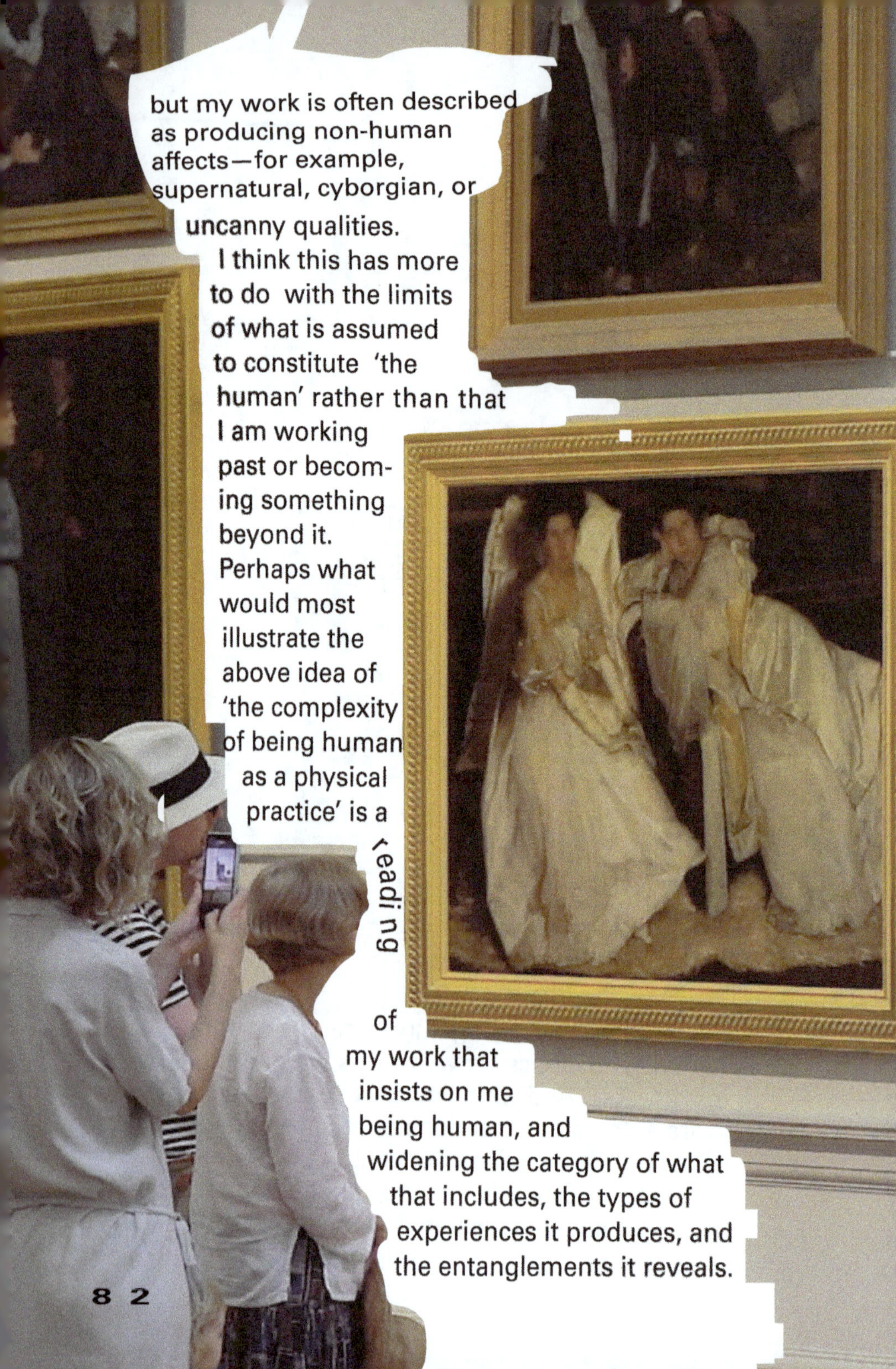
but my work is often described
as producing non-human
affects—for example,
supernatural, cyborgian, or
uncanny qualities.
I think this has more
to do with the limits
of what is assumed
to constitute 'the
human' rather than that
I am working
past or becom-
ing something
beyond it.
Perhaps what
would most
illustrate the
above idea of
'the complexity
of being human
as a physical
practice' is a
reading
of
my work that
insists on me
being human, and
widening the category of what
that includes, the types of
experiences it produces, and
the entanglements it reveals.

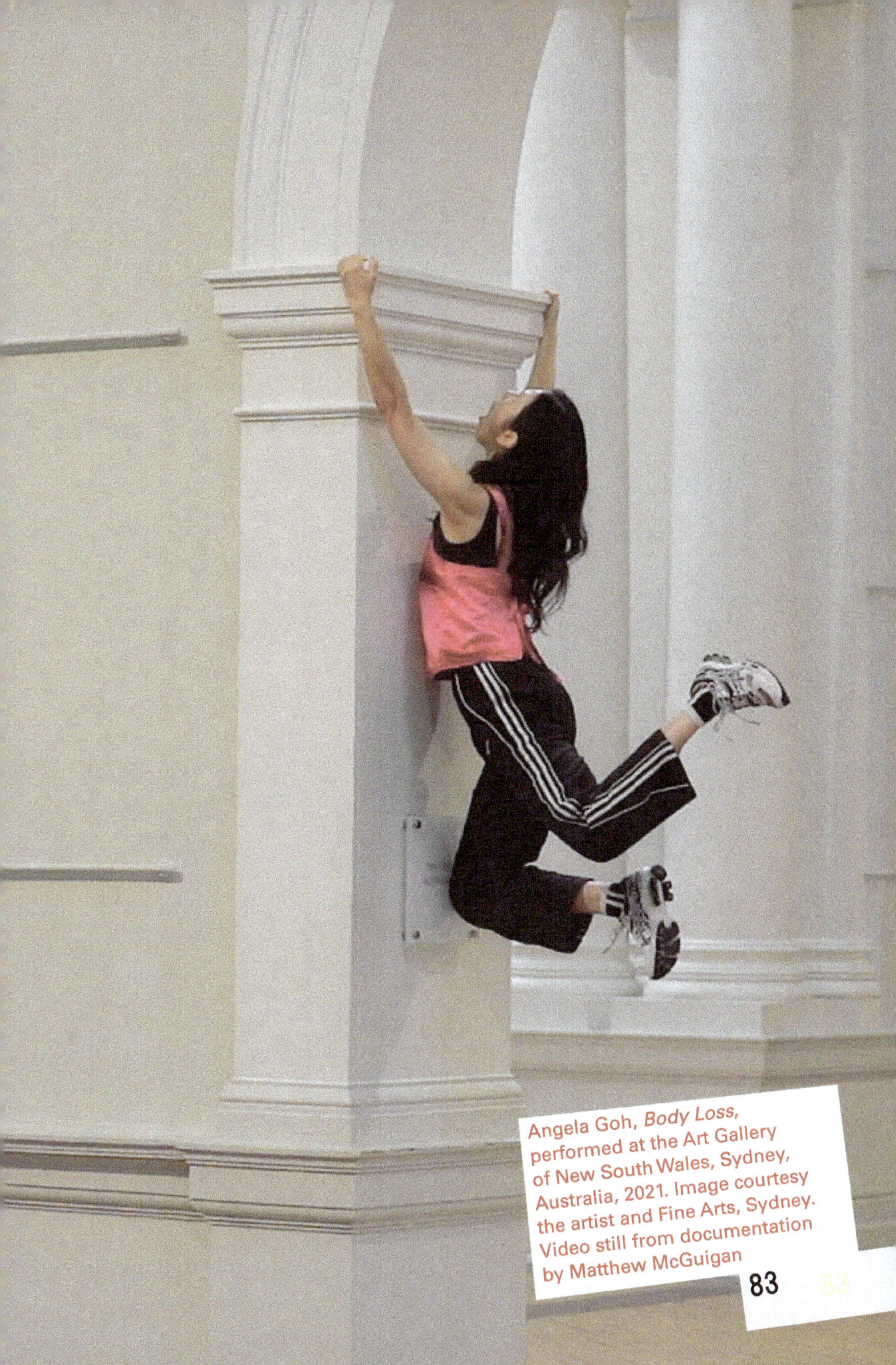

Angela Goh, *Body Loss*, performed at the Art Gallery of New South Wales, Sydney, Australia, 2021. Image courtesy the artist and Fine Arts, Sydney. Video still from documentation by Matthew McGuigan

83

5

Hoping for surplus while imagining repose

Emile Frankel

One always forgets, namely that language, this language which is the instrument of speech, is something material?[1]

In 1994 Sunnyvale, California, a group of hyper-libertarian, techno-utopian, anti-government bodybuilders and nerds sat together in a hot tub at the world's first 'Extropaganza'. Extro 1, as it would be known in those waters, called together believers in 'extropy' (the opposite of entropy). In the pursuit of unending growth, the heightening of all methods of consumption, and a quest for the immortality of the human / market / thought-processing form, Extropians sought a discourse of spending not dissimilar from the description of 'nature' misattributed to Goethe that inspired Freud in his early construction of psychoanalysis:

She tarries, so that one calls out for her; she hurries, so that one never tires of her.

She has neither language nor speech; but she creates tongues and hearts, through which she feels and speaks.[2]

Attached to the side of the Extropaganza jacuzzi, a warning read: 'Please note, some clothing will be required … so as not to shock the neighbours with the sight of our transhuman physiques!' To applause, the software engineer and 'hot blooded capitalist' Mistress Romana arrived dressed as 'The State'. Wearing a specially made leather miniskirt and chain harness and carrying a riding crop in one hand and a leash in the other, 'The State', we are told, had the dog-like figure of her companion, Geoff Dale—'The Taxpayer'—crawling about the dirt in 'mock subjugation'.[3]

Words order themselves around the feeling of pure growth, says the Extropian. A doctrine can itself be extravagant, as if speaking it were spending it. Its members take on new names. The Silicon Valley attorney Tom Bell called himself Tom Morrow. Max T. O'Conner called himself Max More. The Olympian-cum-philosopher F.M. Esfandiary called himself FM-2030.

MIT's AI advisor Marvin Minsky declared Max More the heir to Carl Sagan.[4] More would go on to become the chief executive of Alcor, 'the world's leader in cryonics' (a 'life preserving' company specialising in the cryogenic freezing of human heads). In 1994's hot tub, a member announced: 'immortality is mathematical, not mystical!'[5]

The Club of Life,[6] 'fantastic superabundance', neurocomputers, a memetic approach to selling cryonics, 'Order Without Orderers',[7] and of course Bataille's call towards the lavish expenditure of energy—'it must be spent, willingly or not, gloriously or catastrophically'[8]—united Extropia's doctrine: the seeking out of a financial *moreness* to what it means to be human. Their motto, 'Boundless Expansion, Self-Transformation, Dynamic Optimism, Intelligent Technology, and Spontaneous Order', was very literally yelled as this group of tech CEOs, lawyers, coders and charismatic charlatans clasped their hands together and raised them to the sky.[9] Now was the age for the dynamic optimism of technology, they said. As a very real response to a growing awareness of the environmental consequences of unchained consumption, Extropianism was an antidote to an older philosophical (and cosmic) pessimism[10] that had been building in the sciences and the arts.

In 1893 the palaeontologist Louis Dollo announced his 'law of irreversibility':

An organism never returns exactly to a former state, even if it finds itself placed in conditions of existence identical to those in which it has previously lived.[11]

In 1920, untethering the cyclical time of the Greeks, Rainer Maria Rilke (as translated by Hannah Arendt) wrote:

Here even the mountains only seem to rest under the light of
the stars; they are slowly, secretly devoured by time; nothing
is forever, immortality has fled the world to find an uncertain
abode in the darkness of the human heart that still has the
capacity to remember and to say: forever.[12]

And in 1970, the French biochemist (and thinker of chance)
Jacques Monod condemned a belief in 'the eternal recurrence [of
the] human species':

The ancient covenant is in pieces; man knows at last that he
is alone in the universe's unfeeling immensity, out of which
he emerged only by chance. His destiny is nowhere spelled
out, nor is his duty. The kingdom above or the darkness
below: it is for him to choose.[13]

As species extinction, entropy and omnicide became the
dominant scientific diagnoses of the future, Extropians hoped, as
it were, to turn back the clock—to emphasise something deeply
emancipatory in taking control of the responsibility proffered
by thoughts of finitude. The cosmos, expanding in all directions, was to be directed. Finitude was to be renounced.
As Corey Pein notes, the journal M. More and T. Morrow
started, *Extropy*, 'promoted seafaring secessionism long
before Peter Thiel's Seasteading Institute … It extolled
the subversive potential of digital currencies before
Bitcoin … it denounced, with eerie glee, environmen-
talist, "statists," and "deathist" cryonics critics who
threatened the transhuman future'[14]—and arguably,
for the sake of this essay, the bizarre influence of

Extropy predated the immortalist desire for an endless and always increasing *material* to language.

Those attendees and readers of the journal of the Extropaganza form a direct line to Large Language Modelling.[15] In a blank prompt bar the spores of expenditure flourish alongside the brilliance of *spontaneous order*. The ideology of a gapless plentiful universe, lying in wait for capitalist extraction, bleeds into the ideology of gapless words. Words to be lavishly and catastrophically counted. In Large Language Modelling I see alive and well the seeking out of the efficiency of a lossless compression of meaning against finitude. But I also notice a contradictory philosophy of the endlessly repetitive, of the infinite generation of content—of the problem of surplus in a superabundant future; the problem of seeking stability in an ethos and practice of endless increase.

There's a word Descartes used—'interminate'—which qualifies this feeling of 'no end'. It predates an alienation Lacan picks up.[16] Descartes refuses to label his world 'infinite'. He reserves that denomination for God. Instead, his thought world is without boundaries. In Alexandre Koyré's description:

> **[Descartes'] universe is not infinite (*infinitum*) but 'interminate' (*interminatum*), which means not only that it is boundless and is not terminated by an outside shell, but also that it is not 'terminated' in its constituents, that is, that it utterly lacks precision and strict determination. It never reaches the 'limit' ; it is, in the full sense of the word, *indetermined*. It cannot, therefore, be the object of total and precise knowledge, but only that of a partial and conjectural one.[17]**

'Interminate' then is an expansive 'without end'. Its secondary meaning is a menacing and threatening feeling. The threat of losing an end. The inability to have anything but a partial and imprecise knowledge of the world (and its openness) is expressed by this baleful word. It is a word attentive to the tension (and louring cast) between the unconscious (and coded) labour of finitude and the very sense of no-limit.

The possible interminate 'materiality' of language and the 'quantification' of communication is the concern here. Lacan reminds us that language gained measurability after the invention of Bell's telephone.[18] Language after this moment became culturally inseparable from its energy. A quantity of 'information' now travelled along wires. In his perhaps less-compelling musings on 'telepathy', Freud also wondered if the telephone pointed to a future where we had access to the physicality of formerly unvoiced speech: 'And only think if one could get hold of this physical equivalent of the psychical act!'[19] When communication became principled by its medium, Alexander Graham Bell famously misheard the spirits of the dead in the static of his own copper wires. Lacan reads a different death in this energised sound: the death drive.

Many technologies to record and repeat language were developed during Freud's life. Beyond Freud's fascination with an archaic children's pad, there came the phonograph, the radio, and in the 1890s (before wires could sustain the sound of a human voice) Samuel Morse experimented with transmitting language via energised code. Much has been made of Freud's homeostatic analogy between the anatomic body and the pleasure principle. Lacan instead reads a history of technology through this homeostat.

Asking why Étienne Bonnot de Condillac could not theorise the give-and-take of pleasure in his treatise on the mind, Lacan makes a historicising claim:

**Condillac wasn't deluded. Why, it must be asked, doesn't he give an explicit formulation to the pleasure principle? ···
He didn't have a formula for it because he came before the steam engine. The era of the steam engine, its industrial exploitation, and administrative projects and balance-sheets, were needed, for us to ask the question · what does a**

machine yield?[20]

Prior to the steam engine more came out of the mind than was put in. But after the steam engine, an energetic vision of the mechanical body called into equivalence the restlessness of a clock. There came the sense of an interminance to homeostasis. A compulsion to be 'in-knowledge' of equalness, without end. As Freud grappled with the 'beyond' of his pleasure principle he asked 'what, in terms of energy, is the psyche?'[21]

The fantasy of this question is important today. The Large Modelling machine forces theorists of the inner life to adapt the economy and energies of pleasure to the reconfigured terms of an externalised self-fulfilling imagination. In analysing the relationship between someone who types into a prompt bar and the quality of the machine's generated result, the central tensions of this act respond to the externalised terms of repetition and homeostasis progressed by Large Language Models. The material quality of this coded language (because we search for it) is of importance. Matter in these models repeats itself. If it goes it can be found again. The material can be resuscitated. If the cut is recorded it can be healed again. If anything new is added to this modelled place its addition to the symbolic order produces its own material past. The quantification of speech spurred into action by the telephone reaches its culmination in generative AI: all words receive a value, an energy judgment, which turns language into image, and image into a laboured new discourse tasked with fighting the slow arrow of entropy.

Super Abundance, Spontaneous Order, Order Without Orderers—in reading the production of the fantasy of the interminate through the extropy of Large Language Modelling, the questions 'why more content?', 'why no limits?' are posed against the energy of pleasure and the supposed materiality of language. In the artificial possibility of readable text produced from the complex counting of all prior catalogued words, prompt bars and their mechanisms expose new ways to read Freud's dialectical terms: excitation and stability; pleasure and reality; the compulsion to repeat the present or the compulsion towards a restitutive beginning.

These terms find new consequence in today's AI abundance of word-combination. Clearly words are decreasing as they increase. Surplus is misrecognised as repose. The great proliferation of modelled content marks the beginning of the endless generation of not quite what we want. Content only marginally good enough, acceptable enough to warrant consumption, but imperfect enough to keep us wanting more. The lure of such generation must surely be found at once in the promise of endlessly up-ticking growth—endless surplus—but also in the flatline such an oxymoron proposes. The number goes up but its increase approaches zero. Is this or is this not the Extropaganza?

1   Jacques Lacan, *The Seminar of Jacques Lacan Book II: the Ego in Freud's Theory and in the Technique of Psychoanalysis 1954–55*, WW Norton, 1997, p. 82..^l

2   See Friedrich A Kittler, *Discourse Networks, 1800 / 1900*, Stanford, Stanford University Press, 1990, p. 25. The quote was incorrectly attributed to Goethe, but was in fact from Georg Christoph Tobler's essay 'Die Natur', 1783, which was written as a result of the author's many long conversations with Goethe..^l

3   Description curtesy of attendee Ed Regis, 'Meet the Extropians', *Wired*, 1 October 1994, https://www.wired.com/1994/10/extropians/..^l

4   Corey Pein, 'Everybody Freeze!: The Extropians Want Your Body', *The Baffler*, no. 30, 2016, p. 90..^l

5   The statement was attributed to Mike Perry. Regis, 'Meet the Extropians'..^l

6   The Club of Life was founded in 1982 as a response to The Club of Rome. See Lyndon H. La Rouche, Jr, *There are No Limits to Growth*, New York: The New Benjamin Franklin House Publishing Company, 1983..^l

7   H. Keith Henson and Arel Lucas, 'A Memetic Approach to Selling Cryonics', Extro 7..^l

8   Georges Bataille, *The Accursed Share: Volume 1: Consumption*, trans. Robert Hurley, Zone Books, 1988, p. 21..^l

9   The 'Extropian Handshake' is described in Regis, 'Meet the Extropians'..^l

10   See Eugene Thacker, *Cosmic Pessimism*, Minnesota, University of Minnesota Press, 2015..^l

11   S.J. Gould, 'Dollo on Dollo's Law: Irreversibility and the Status of Evolution-

ary Laws', *Journal of the History of Biology*, no. 2, 1970, pp. 189–212..^|

12   Rainer Maria Rilke, *Aus dem Nachlass des Grafen C.W.*, quoted and translated by Hannah Arendt in *Between Past and Future: Six Exercises in Political Thought*, New York, The Viking Press, 1961, p. 44..^|

13   Jacques Monod, *Chance and Necessity*, trans. A. Wainhouse, London, Collins, 1971, p. 167. Quoted by Thomas Moynihan in *X-Risk: How Humanity Discovered Its Own Extinction*, Mass., MIT Press, 2021, p. 125..^|

14   Pein, 'Everybody Freeze!', p. 90..^|

15   See Henny Ge Wichers, 'TESCREAL', *Generative AI*, 12 May 2023, https://generativeai.pub/tescreal-b271de909133; and Max More's substack, 'Extropic Thoughts', accessed 8 October 2023, https://maxmore.substack.com/?utm_source=substack&utm_medium=web&utm_content=comment_metadata.^|

16   See Seminar-IX: alienation cannot be theorised before Descartes. And Seminar-XI: 'the Freudian field was possible only a certain time after the emergence of the Cartesian subject.', p. 47..^|

17   Alexandre Koyré, *From the Closed World to the Infinite Universe*, Baltimore, Johns Hopkins University Press, 1957, p. 8..^|

18   Lacan, Seminar II, p. 82..^|

19   Freud, 'Dreams and Occultism', SE:22, p. 55..^|

20   Lacan, Seminar II, p. 61..^|

21   Ibid., p. 75..^|

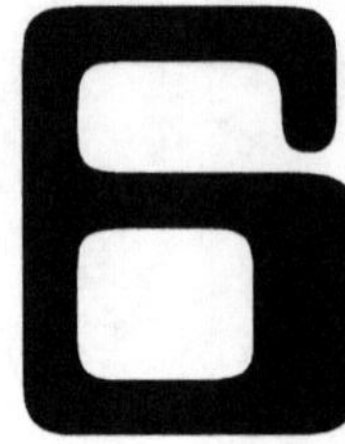

WAKING LIFE: THE DREAMWORK MODEL

Thomas William Smith
with Jon Watts

It is the year 2422, and what you are about to hear is the story of humanity's gainful optimisation, told in the style of what was once known as a sleep story, a curious form once popular in the 21st century. Of course in our time, no one ever sleeps, but in ages past, sleep was common and essential. In the late 21st century, sleep and productivity came into conflict. As workers became increasingly alienated from the products of their labour, and communication technologies were integrated into all areas of life, sleep became fleeting and elusive. To remedy this problem, many aspiring sleepers turned to sleep stories voiced by well-known celebrities. Companies with names like Headspace, Calm, Moshi Twilight and Sleep Cycle Alarm transformed information, concepts, places and ideas into databases of adult lullabies. My use of this format aims to convey something of the past, to help listeners understand the lost world of anaesthesia, lethargy and somnolence in which our forebears lived and died.

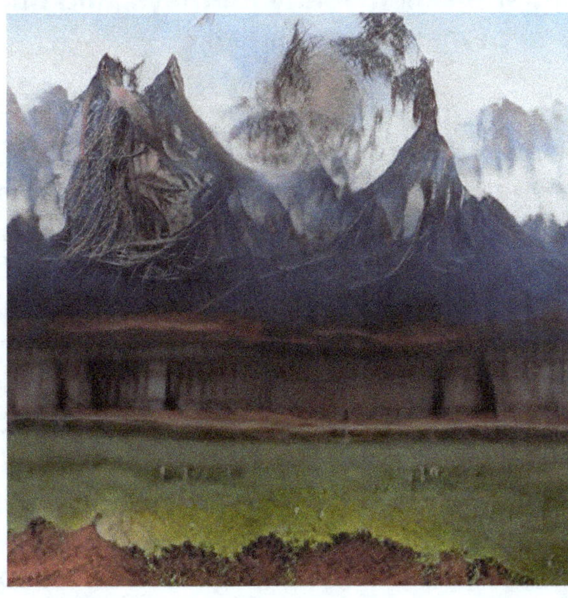

Let's begin with a brief overview of research leading to the elimination of sleep.

In the year 2310, the Defense Advanced Research Projects Agency, also known as DARPA, finally achieved a longstanding goal, the creation of sleepless soldiers. Over the preceding century, research was advanced in several directions. Initially, drugs such as caffeine, amphetamines and modafanil were studied. Taken in carefully controlled doses, these drugs could counter the effects of sleep deprivation for several days. Later, DARPA began trialling a new experimental class of drugs known as ampakines. Acting upon glutamatergic receptors in the brain, ampakines allowed even longer periods of wakefulness. In the hope that their adaptations might be transferrable to humans, various animals were also studied extensively. Elephants, giraffes and other large grazing animals sleep for only two to four hours a day. Whales remain awake for weeks at a time to nurse vulnerable young. Migratory birds remain awake for several days while completing their yearly journeys, and bullfrogs are thought to perhaps never sleep.

Many birds and marine mammals practice uni-hemispherical sleep, in which one hemisphere of the brain sleeps, while the other remains active, maintaining vigilance and basic situational awareness. The possibility that human beings might also sleep in this manner was explored for many years. Through a combination of pharmaceuticals, gene therapy, and advanced meditation techniques developed in the 21st century's wellness industry, a minority of successes were recorded. Experiments in this area included isolating the genes regulating sleep in Cape Fur Seals, allowing them to be introduced to human neurons in the basal forebrain, anterior hypothalamus, midbrain and brain stem. A series

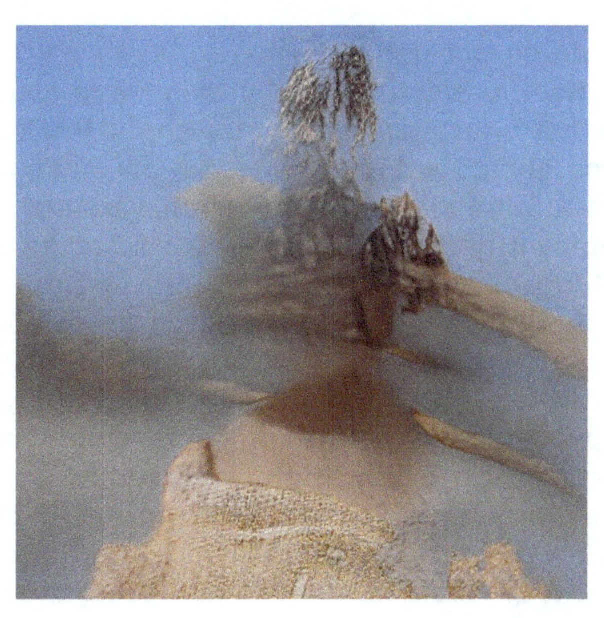

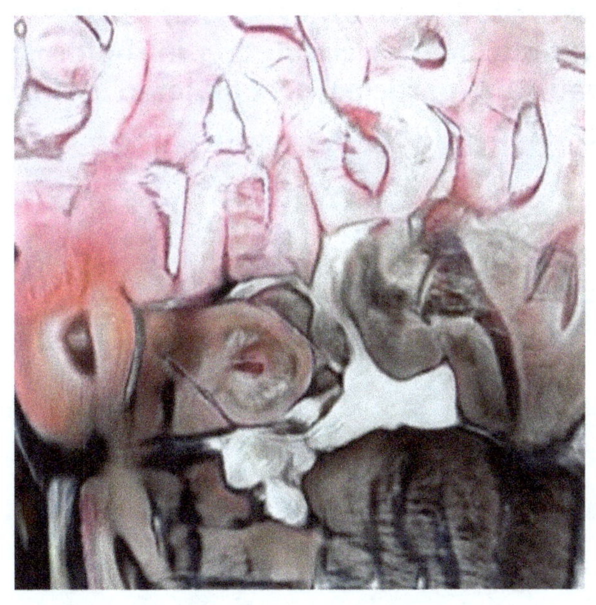

of drugs consisting of hypothalamic growth hormone, cortisol, prolactin and thyroxine was also developed to further alter these brain structures. DARPA scientists ultimately judged uni-hemispherical sleep to be of limited use to humans, as test subjects were not able to perform sufficiently varied or sophisticated tasks while sleeping. Additionally, the extraordinary mental discipline required made it unsuitable for commercialisation.

However, techniques aimed at eliminating sleep entirely proved more scalable and profitable. One of the most important breakthroughs involved redesigning the process through which harmful byproducts of cognition are removed from the cerebrospinal and interstitial fluid in which the brain is suspended. During the process of cognition, neurons pass chemical and electrical signals to one another across synapses. Although this process is very efficient, around 0.1% of neurotransmitters are not absorbed by synaptic receptors, instead remaining in the synaptic cleft or circulating in the brain fluids. As these wayward particles accumulated during waking hours,

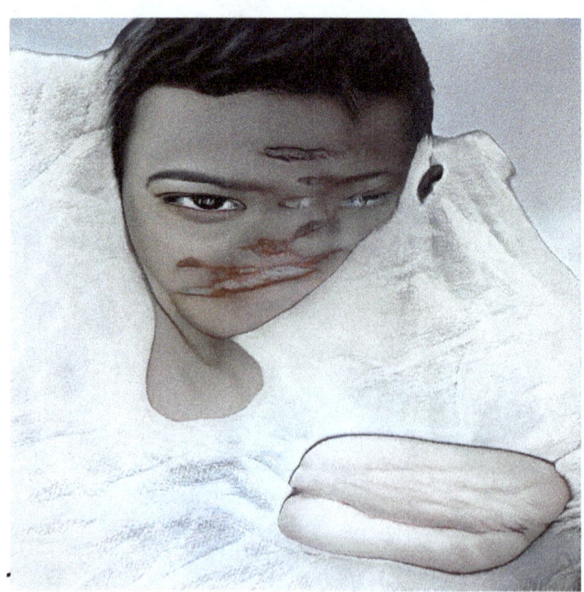

the efficiency of synaptic activity was reduced, and cognition impaired. During non rapid eye movement sleep, convective flow through the brain was greatly increased, flushing neurotransmitters and other byproducts of neuronal metabolism from the brain. It was discovered that a combination of plasma proteins, globular proteins, albumin, sodium and various chlorides administered directly to the choroid plexus would slightly alter the chemistry of brain fluids and promote the production of microglia cells, which clear cellular debris from the brain. Surplus neurotransmitters and other toxins could now be removed on an accelerated and ongoing basis. This also drastically reduced build up of the beta-am-

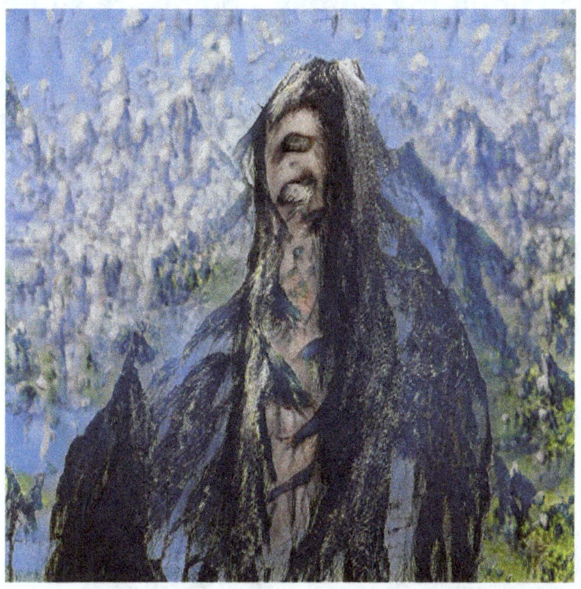

yloid 42 plaque that once caused Alzheimer's disease, effectively eliminating the condition.

Sleep research followed this pattern for over a century. As research progressed, incremental gains in the efficacy of pharmaceutical intervention and gene therapy allowed for the simulation of all sleep's beneficial functions, and the prevention of all sleep deprivation's detrimental effects. Although these symptoms are now historical, sleep deprivation was once associated with drastic reductions in cognitive capacity, immune response, behavioural control, bodily co-ordination, reaction times and short-term memory function; along with increases in body mass index, hypertension, chronic pain, inflammation, mood swings and hallucinations. Long-term effects of sleep deprivation once included alteration of the human DNA transcriptome, and greater instance of anxiety, substance abuse, heart disease, Post Traumatic Stress Disorder and Alzheimer's disease. The effects of sleep upon the human body were so numerous and diverse that in order to simulate its effects, 557 discrete compounds, hormones, catalysts and peptides were developed and administered in precise quantities via a time controlled implant.

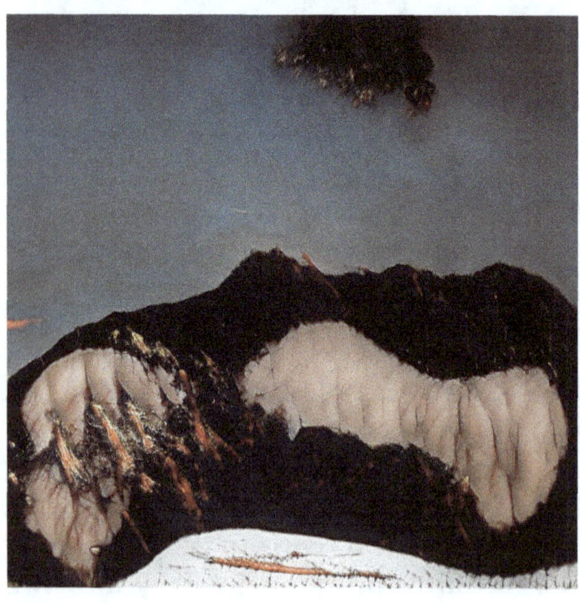

Initially, these techniques were restricted to the military, and were subject to the utmost secrecy in order to maintain tactical advantage. Like many of the technological solutions now taken for granted, the military origins of sleep elimination have been largely forgotten. As the system of tenders and contracts that constituted the military industrial complex waned in the late 23rd century, DARPA diversified its client base in the private sector. It was soon recognised by various entrepreneurs that the elimination of sleep would provide the most rapid increases in productivity since the invention of industrial machinery in the 19th century. The commercialisation of the vast array of pharmaceuticals necessary to achieve sleeplessness helped the pharmaceutical industry grow by approximately 660% as the technique became culturally normalised, and eventually universal among the nations of the global north in 2391.

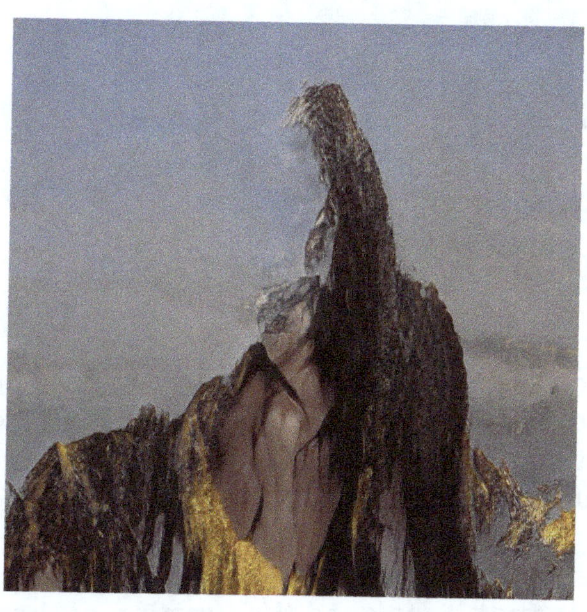

Although most labour has been automated, there remains a suite of tasks that are still performed by humans, whether for practical or political reasons. These include garment production, tree surgery, conveyancing, personal training, mail delivery, and certain hospitality roles, which are performed by citizens excluded from income redistribution as punishment for various crimes. It is however in the area of cultural production that the effects of automation and the elimination of sleep have been most varied and complex: communication, writing, research, speaking, artistic ideation, musical composition and so on became the basis for economic growth in the over-developed world. The products

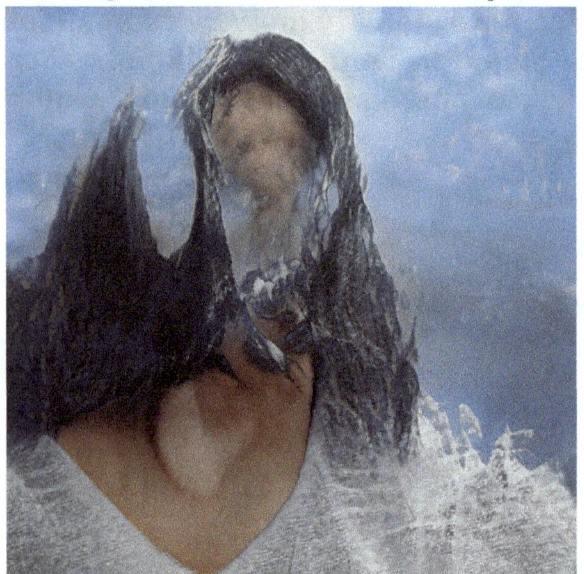

of intellectual and artistic labour are now produced and consumed 24 hours a day by the majority of citizens; a constant stream of intellectual, cognitive and aesthetic labour that is both enjoyable and essential.

The act of consuming media became necessary for survival in that it replaced the function of dreaming. Of all the rituals, happenings and experiences that once accompanied sleep, dreaming was perhaps the most culturally and psychologically important. Research conducted in the 21st century confirmed what Sigmund Freud partially explained over 500 years ago in 1899. Dreaming had once enabled humans to process emotions, memories and experiences. Dreams were the means through which raw experiences were transformed into memories and incorporated into a coherent personality and sense of self. The visualised world of disembodied objects, people, places and desires that once constituted our dreams served to reinforce valuable synaptic connections, and weaken others. Through this process of unconscious cognition, the happenings of the day—along with older memories—were examined, categorised, reinforced, or forgotten. This 'dreamwork'—as Freud once referred to it—is no longer performed internally in a haphazard manner. It is now performed immanently, as a function of media production and consumption.

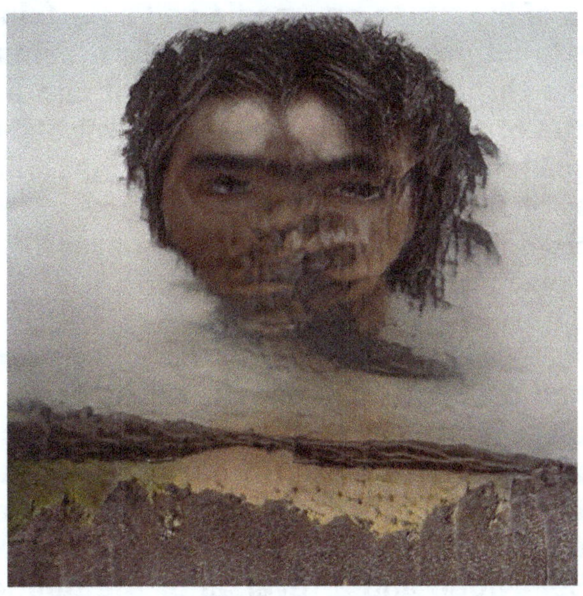

To retain some of the richness of human experience and a modicum of diversity among personalities, it was necessary to outsource the production and consumption of dreams to an advanced neural network trained on video, images and sounds captured throughout the day by the general population. Everyone must contribute to the 'Automated Dreamwork Dataset'; an ever-expanding archive of memories now estimated to contain over 5000 zettabytes of personal recordings and other data. The 'Dreamwork Model', a neural network that draws on this media to generate salubrious and socially advantageous dream scenarios, has been training on this dataset for over 30 years. The visual material you are seeing now is a selection of early dreamwork experiments. The consumption of dreamwork was incorporated into the distribution processes of the entertainment industry, ensuring it would become a collective endeavour performed through the act of everyday participation in culture. Of course, the world of literature, art, media and entertainment has always functioned as a repository for dreams and stories, filtered through the political, social and economic apparatus of the day. However, the symbolic dreamworld of ages past is now a literal world that constitutes the realm of the social. Dreaming is now a technological supplement, and participation is necessary for survival; in our time, we all dream as one.

The automation of dreamwork had a number of social and psychological effects. In the same way sleep gained secondary therapeutic characteristics not present at its initial evolution, the faculty of sleeplessness began to evolve its own secondary functions. These include the elimination of mood disorders and anti-social behaviour, and a reversal of the political and discursive polarisation that characterised the public sphere of the 21st and 22nd centuries. A series of psychological traits once associated with Cognitive Behavioural Therapy became universal; neuroses, anxiety disorders and paranoid ideation were greatly reduced. The Dreamwork Model was trained to omit certain types of recorded memories from its dream creations, de-activating synaptic chains associated with trauma related psychiatric disorders. Personalities are no longer a mélange of our own experiences filtered through our sleeping minds. They are now based on a set of predetermined principles and goals established by an agency called the Bureau of Interpersonal Wellness, under whose guidance, the cognitive processing that produces memory and personality were standardised. The benefits of automated dreamwork have not been questioned openly for over 30 years, since the system came online in the year 2391. The genetic aspects of the sleep elimination process have now become heritable and irreversible. Individuals who attempt to disconnect from the automated dreamworld

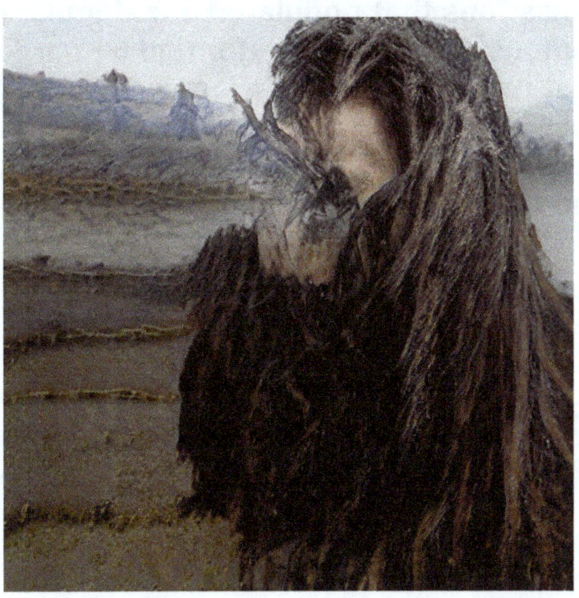

suffer rapid degradation of cognitive abilities, and an acute form of psychosis known as Core Principle Abandonment Syndrome. Though of course, very few among us desire a return to the previous world of slumber and indolence.

What you have heard today is a short introduction; a mere cursory outline of the innumerable societal changes that have accompanied sleeplessness in humans. To accurately describe these changes would require a dream story of unprecedented length and complexity. Perhaps one day, this story will be told in all its nuance, tragedy, and of course, triumph. Until then, let's reflect upon the remarkable advances that have so greatly enhanced our living and functioning. If you were a fatigued citizen of the 21st century, solemnly listening to this story in the hope of losing consciousness, you would by now be asleep. You would be cast helplessly adrift amongst the symbols, fragments, displacements, dramatisations and condensations that populated our singular dreams. The perilous dreamwork performed by our former selves kept us marooned upon lonely islands of unregulated cognition. Thankfully, automated dreamwork has now become the very substance of our gainfully optimised and permanent waking life.

SCRIPT & MUSIC Thomas William Smith
IMAGES Jon Watts & Thomas William Smith
VOICE Dave Pettitt Commissioned by
Anabelle Lacroix and
Liquid Architecture for
the Freedom of Sleep program

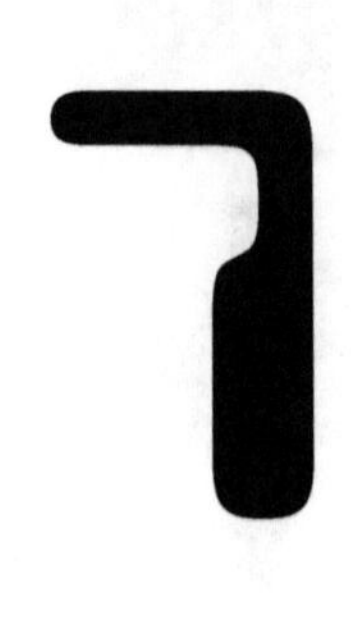

3 Degrees of Freedom

Ying Ang & Ling Ang

Commissioned by Photo Australia for PHOTO 2022
International Festival of Photography
Courtesy the artists

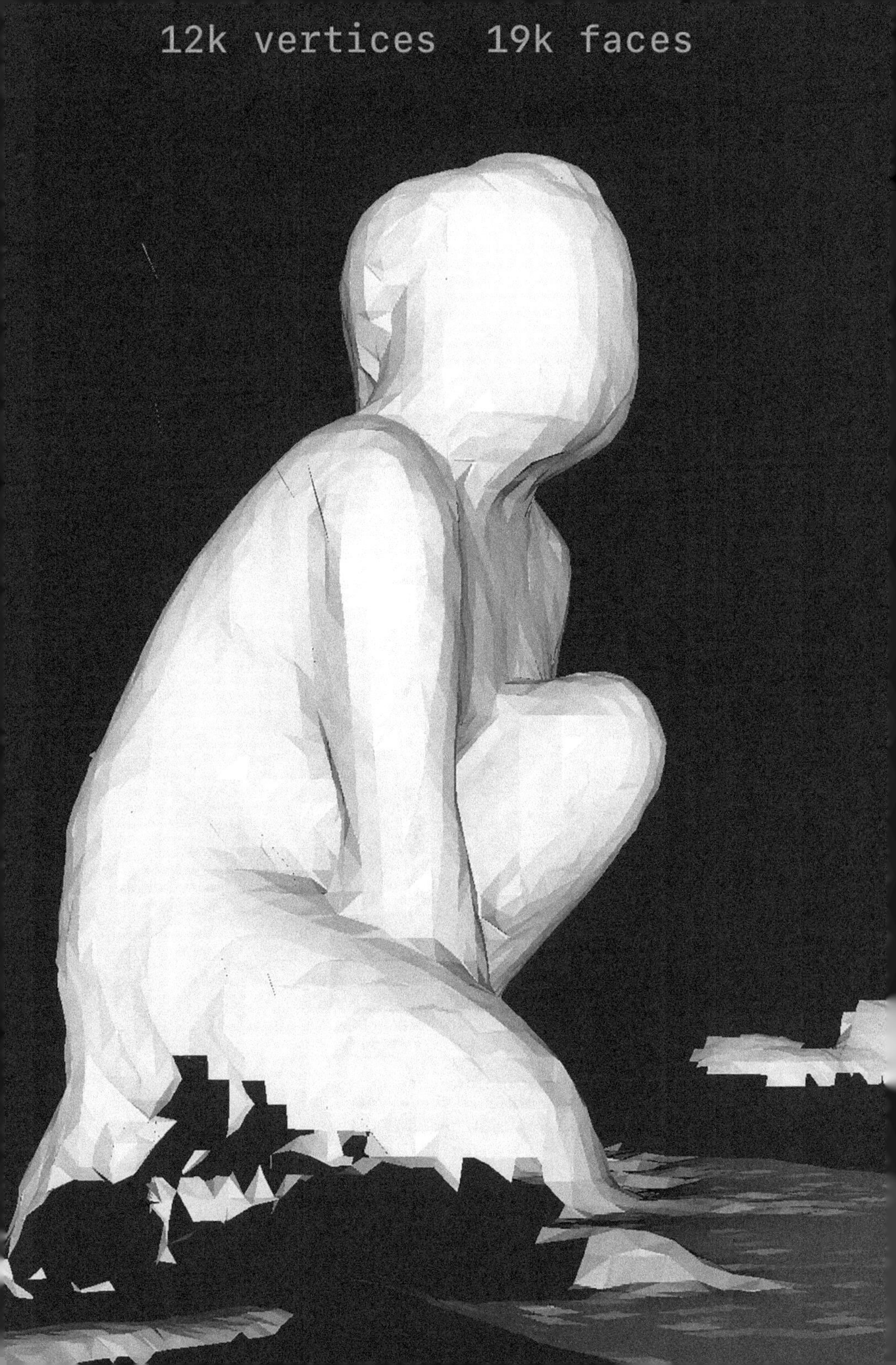

12k vertices 19k faces

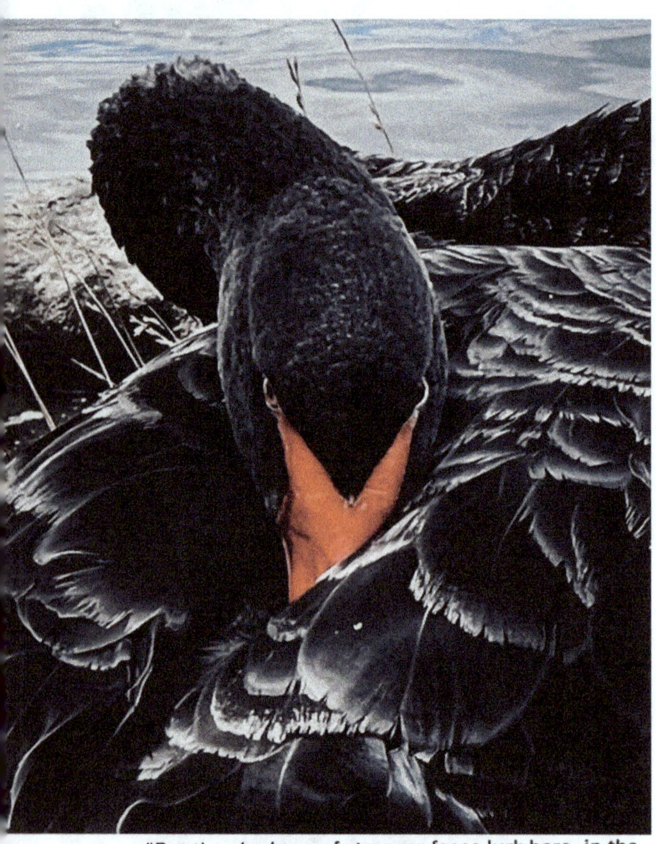

"But the shadows of stranger faces lurk here, in the Instagram Face's figural hybridity, racial ambiguity, clay-like sculptural thingness, animal features, blank sublimity, eerie layeredness, and, of course, in its origin point and favored playground: the internet."
–from "Stranger Faces" by Namwali Serpell

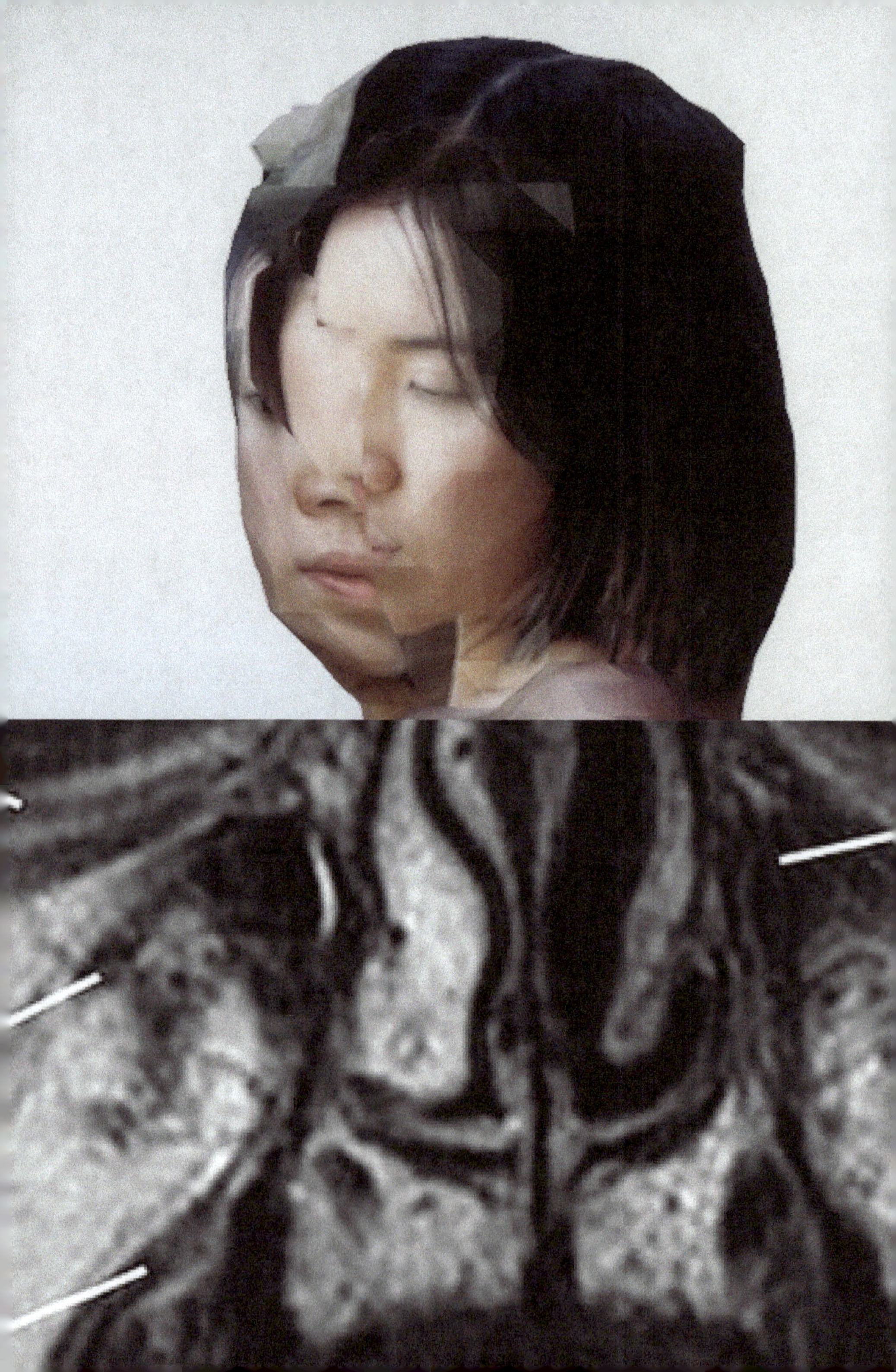

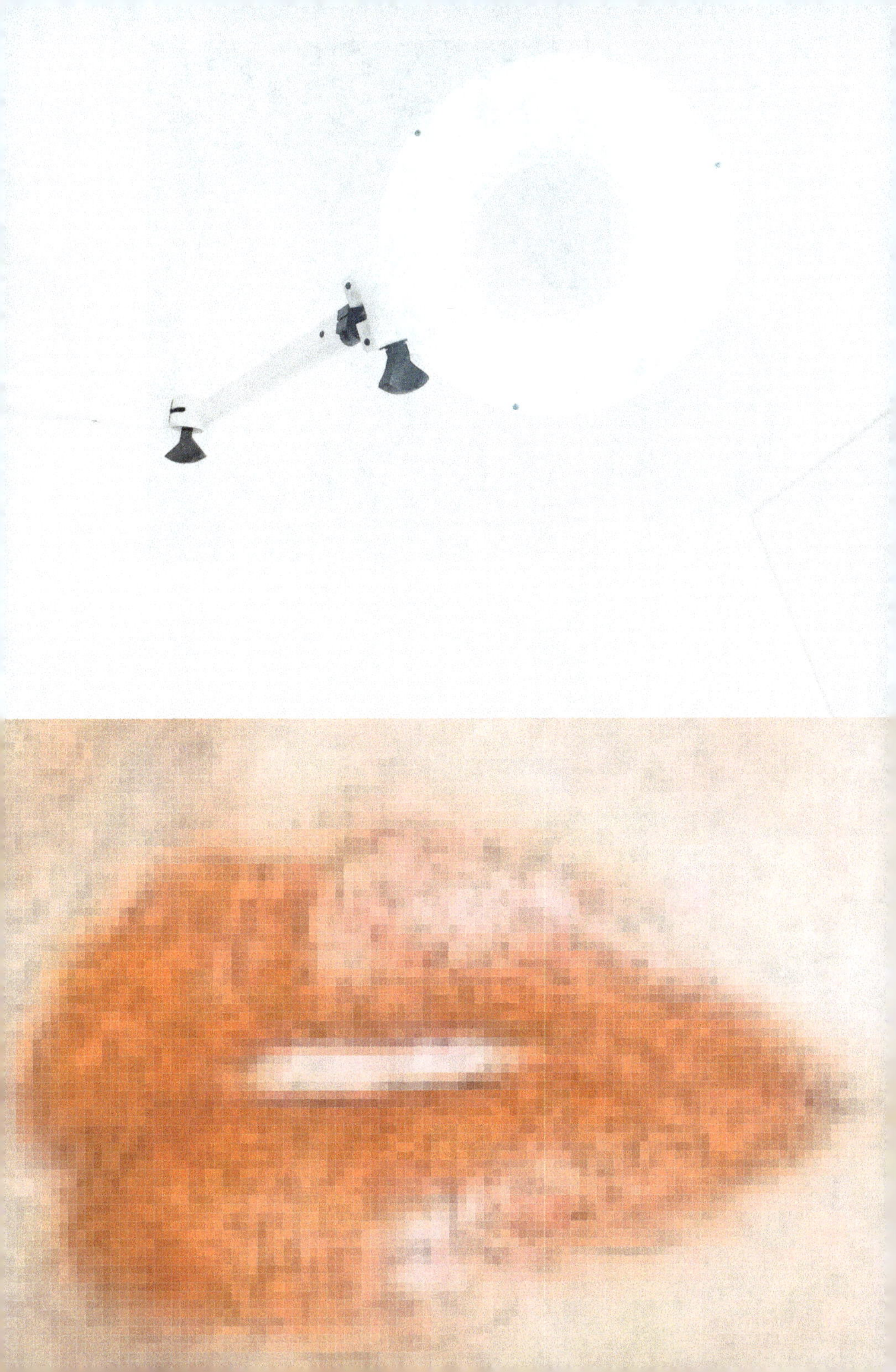

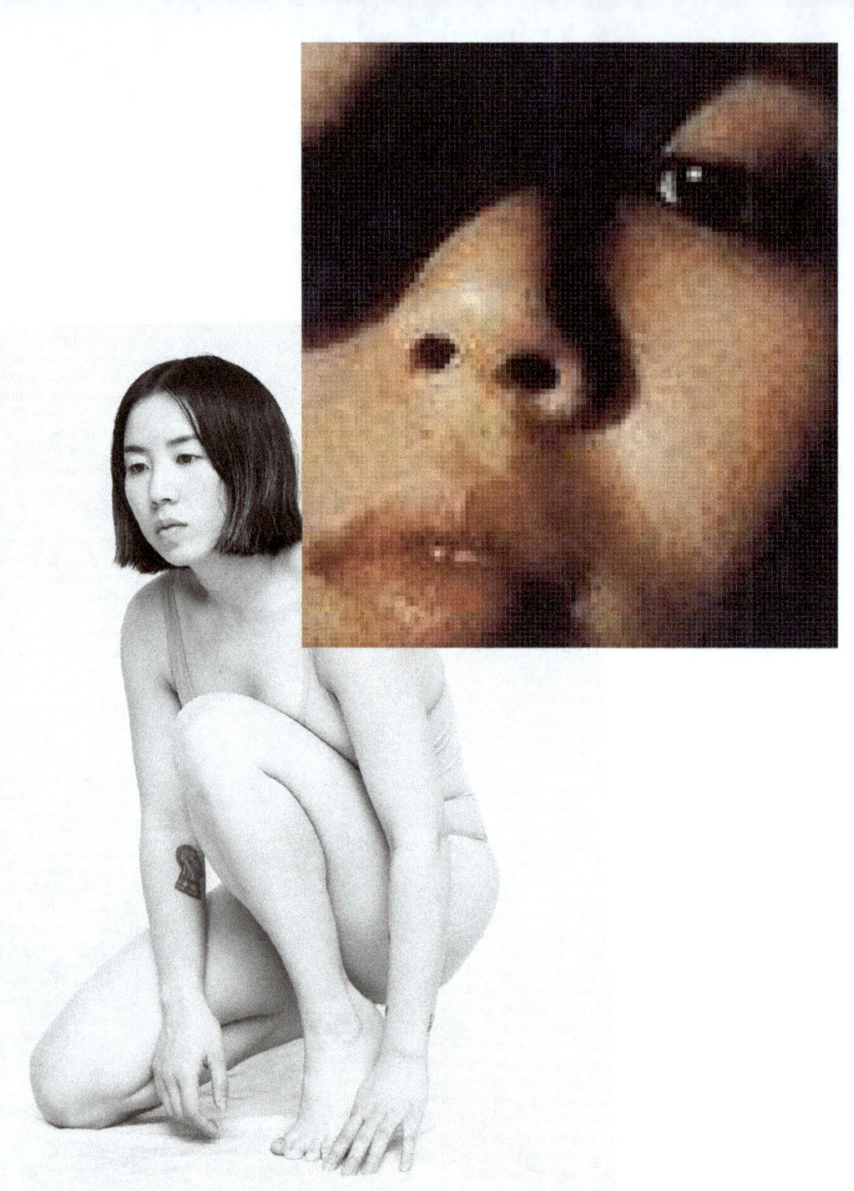

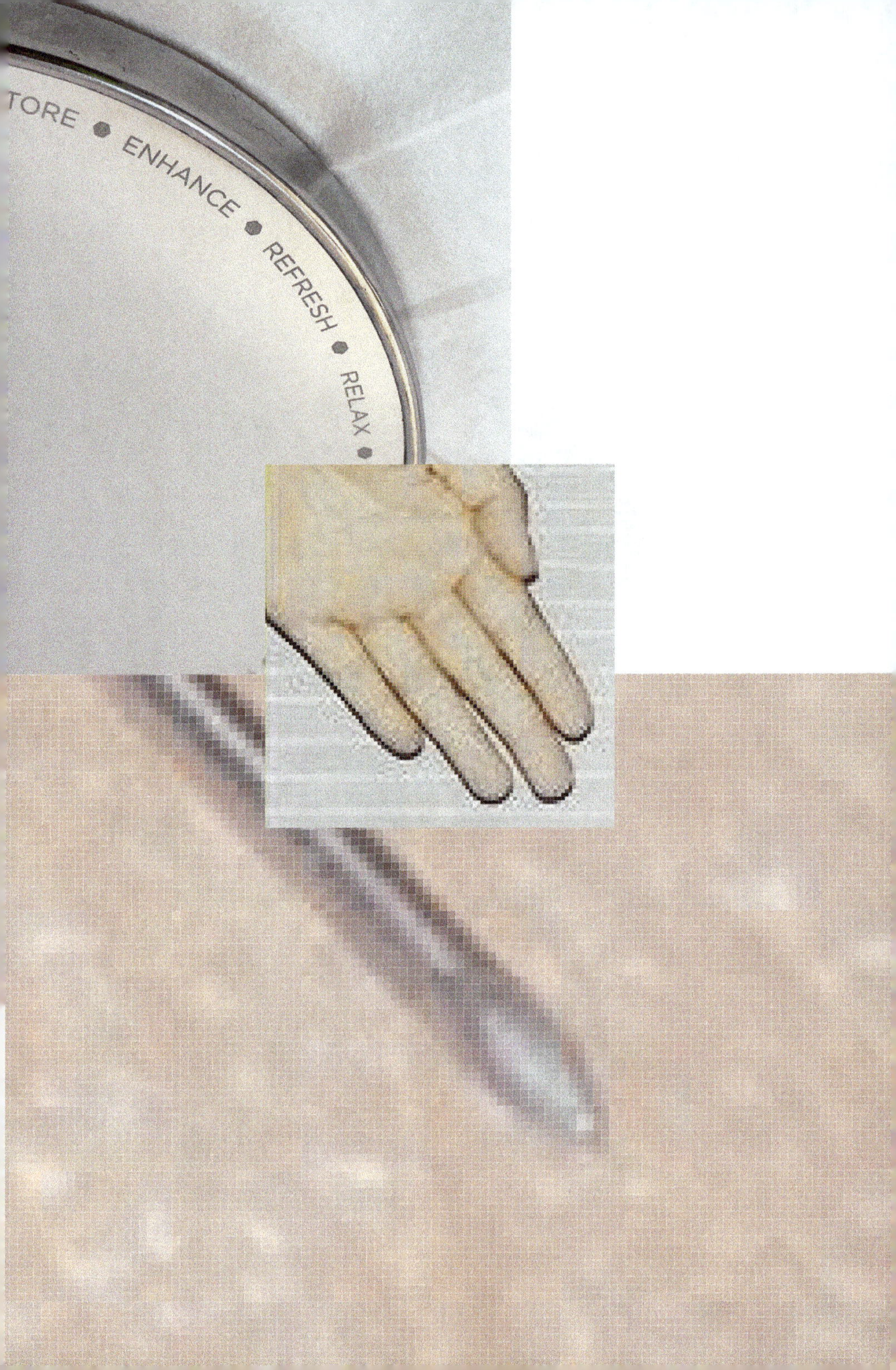

TORE
ENHANCE
REFRESH
RELAX

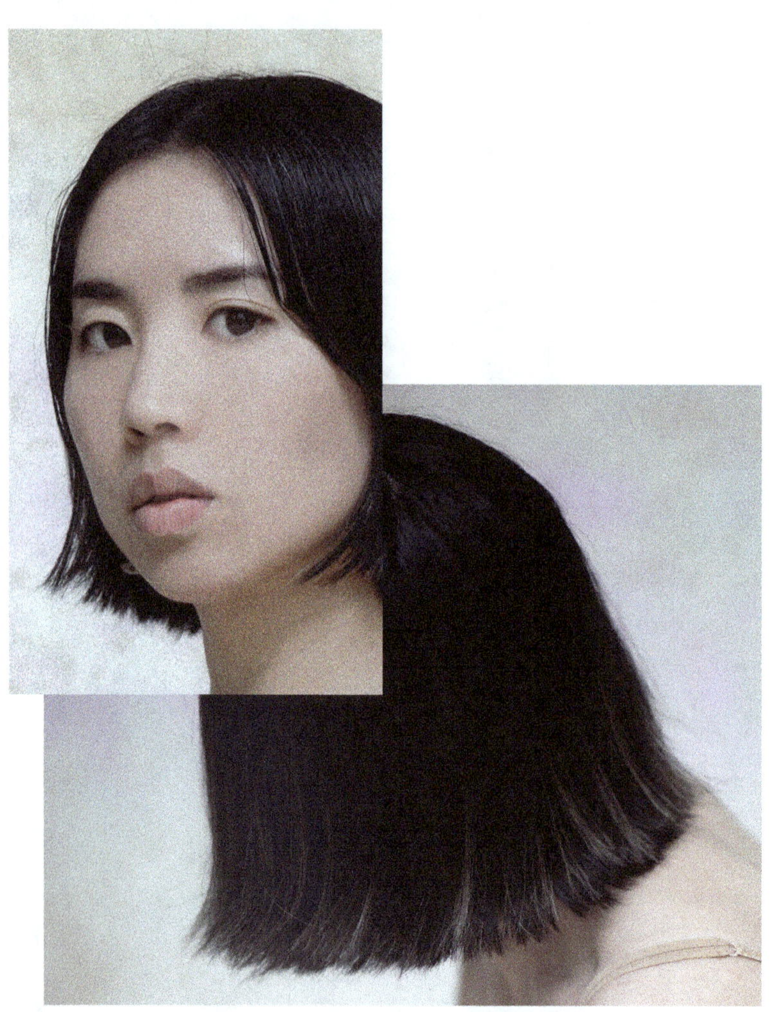

In the current era of the digital revolution, we find ourselves at the cusp of the Self as the dominant currency in the social economy. The increasing power behind big data and algorithmic manipulations has led to the rise of the eternally optimising self locked in the pursuit of increasing influence and ultimately social power.

We have exalted the virtual human icon that biological humans are increasingly using to orient their striving. This work is a comment on self-surveillance, self-optimisation and the expanding state of self-idolatry that is a product of the algorithmic mechanisms feeding the myth of our identities. We are being moulded by corporate interest to mine our sense of self and to continuously invent and reinvent in a cycle as vicious and cannibalistic as a snake eating its own tail. With every contortion, there is a transaction and a promise of being one step closer to perfection.

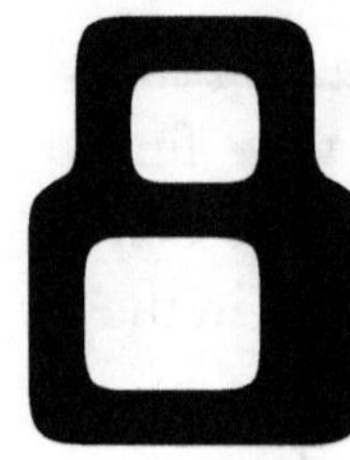

The Disembodied Satisfier

Luara Karlson-Carp

'Everyone is female and everyone hates it'—this statement forms the central thesis of Andrea Long Chu's counter-canonical 2019 text, where being female is defined by possessing the desire, at a fundamental level, for someone to do your desiring for you. This formulation seems to figure desire, in essence, as a wish to be free *from* desire, or at least to offload or even offshore desire onto a desiring other, a *prosthetic* desirer, one willing to do the dirty work of wanting things in a compromised world. *Tell me what to do, Daddy*, is the dirtiest anyone gets for Chu.

If the curtain of the human condition is pulled back to reveal a *universal* squeamishness about desiring for oneself, with no rift nor rupture nor difference in this imagined field of desirers, are we still even talking about desire? In Chu's thesis, desire has one sex only— the female sex—now positioned as the universal (*everyone* is female). This formulation would nevertheless seem to necessitate a second term, a someone (though perhaps more accurately a something) who would desire *for* a female, though within the logical terms of Chu's proposition this something could not itself have a sex. Chu's claim, taking up but pushing far beyond the queer-theoretical zeit-geist, defies the long(ish) historical arc of binary gender doxa in which desire has been figured as hetero-desire by definition. Chu takes up half the psychoanalytic paragon of this view in which desire is figured as *the* force that differentiates the sexes, where, in its most reductive form, 'the man' actively desires, and 'the woman' wants to be desired. But Chu drops the other masculine half, and so instead claims that desire is the most gender-neutralis-ing force of all. Desire, as universally female, negates sex. But what would universal femaleness even *mean*? Where is the second term against which the specificity of this 'femaleness' would determine itself? If the object of desire is the one willing to desire something more determinant than simply someone else's desire, and yet we're all universally, and femininely, incapable of this kind of desire, if God remains dead but we still can't quite let go … What is this second term? Where might it be glimpsed? Who is doing the work of desire?

My claim is that Chu's thesis shares the presupposition of the fundamentally sexuately neutralising function of desire with algorithmic technology. Whilst algorithms may encode gender as a key data point alongside other finely grained 'intersectional' identity markers—age, race, income, vibes—that our digital footprint flags to Big Daddy Mainframe, the *form* of desire the algorithm assumes of its users *is always the same*. For the algorithm, 'users' are all assumed to desire in the same way, according to the same logic. There is no sexual difference at the level of desire for this form of techne; while the identities AI makes footprints of are singular with respect to which products each subject wants, at the most basic level, the level of *how* they want, the subject AI addresses is universal. The content of our feeds or the advertisements that punctuate our scrolling may be different, but the mechanism of techno-seduction is not—the advertising feedback loop tells us what we desire, what's cool, what products would allow us to ascend closer to the ideal our data imputes for us, *equally*. It is in this sense I think that the subject of AI and the subject of Chu's thesis are one and the same: the subject AI produces us as is a 'Chuian' *female* subject, a subject who is tantalised by the promise of offloading the dirty labour of their desire onto someone, or, more accurately today, *something* else—this something else being the code canalising the app, the algorithm designating the feed, the interface smoothing the experience, and the fabric of economic relations overdetermining them all.

Recently, at the university where I teach, there has been some buzzy excitement about new software that allows students to annotate texts together, which has been given the inspired moniker 'Annotate'. It will supposedly enable better 'inter-cohort engagement' during class. I am in a meeting where all in attendance are in various states of dizzy at the prospects such an innovative techy intervention might afford. I make the mental meme: we already have 'annotate' at home; it's called *talking to each other in class* about the assigned texts. But to say so here and now in the meeting would be to commit the highly anti-collegiate sin of talking right past the fantasy at work: those around the table dream that the inscriptive power of this software might possibly reignite the long-lost—and much mourned by university tutors—student desire to engage with texts. I get the sense that my colleagues in this meeting imagine—hopefully, libidinally—that this software might itself become the prosthetic desire for a student's own textual desire. 'Annotate' might do their desiring for them.

Later that night, as I lie scrolling in bed (being Female), I am advertised yet another AI personal assistant. In the ad, a listless and very hot autist bemoans the difficulties of organising their subsistence tasks in the morning—brushing teeth, folding clothes, making breakfast. Apparently the AI 'assistant' will rationalise their schedule such that they need not suffer from executive function dilemmas ever again, ordering their most basic tasks into a seamless, eminently executable procedure. I think about the times I've struggled with this basic level of getting it together—it certainly happens, but the app presents this all-too-human experience of ineptitude in extreme abstraction, reified as symptoms divorced from context and cause. The times I can't perform basic functions coincide either with a hangover or those periods in my life when I cannot bear to publicly desire, where the idea of having to appear before the gaze of others, ruddy with the muck and gall of wanting things, probably things I feel I haven't been very good at getting, is surely disgustingly obvious and odious to all who behold me, and scattered malaise becomes the only defence. When my very apprehension of my own desire, paltry and tepid as it may be, repulses me, I'm plunged into deep ravines of avoidant, rapacious, almost gluttonous sloth and 'executive' disability.

Whilst not wanting to completely collapse my own non-autistic experience with those who have had transformative diagnoses, and certainly not those with low-functioning autism, I am certain that perhaps even a majority of the people who are being advertised to and seduced by this ad are not so different to me—most probably eminently capable of getting a high-functioning adult autism diagnosis, but perhaps more simply and more accurately suffering with a chronic case of fucking it up in a fucked up world and feeling fucked up about it. The ad only works—and likewise, the very possibility of most of the new expansive diagnostic criteria comprising the autism boom today—through a severance of the symptoms' relationship to desire. The condition supposedly has nothing to do with you as situated within a particular world, a particular economy, a particular culture within which you struggle to realise and find and even bear to admit to that which you might desperately want, about the impossibilities that criss-cross human experience differently but ineliminably; no no, it's about your developmentally disordered biochemistry, your congenitally wonky nervous system, your predilection to mask.

Abstracting symptom from world, the apps (and related diagnoses) these kinds of ads promote are able to parade themselves as magic bullets, offering up a tantalising substitution for desire in the form of techy large language model 'workarounds' for the problem of desire itself. If our schedules are planned by zeroes and ones, and our agency milled through a prosthetic, machinic will, maybe it will become irrelevant that we can't find a place for our desire in a world that seems intent on eradicating sociality, let alone housing; perhaps it won't matter that through demoralisation our desire has atrophied to the extent we can't even wield it against the crippling tides of anxiety and quick-fix compulsion that take the place our desire might have once had, if it ever had the chance. Perhaps the very smart machine will do the work of desiring this life that I cannot bring myself to want.

Technology has long been understood in terms of prosthesis— our Smart Phones are as much a prosthetic affordance as a plastic limb, a set of glasses, a hammer. For the late philosopher of technology Bernard Stiegler, technology is not merely about a means to an end, the right tool for the job—technics *is* the 'prosthesis of the human'. It is the ceaseless process augmenting our capacities through the production of their artificial extension and enhancement. Technological prosthetics—which for Stiegler include techniques as elementary as writing all the way up to complex machines and AI systems—constitute the 'exteriorisation', the making-exterior, of our own knowledge and know-how into material objects. And this has existential implications for Stiegler—the human, for him, is not defined by some internal capacity such as consciousness, but by our technically prosthetic relationship with matter, where the human, the prosthetic and the wider material world are locked into continuous mutual transformation. Perhaps, then, we could say that AI is the technical coming of an *immaterial* prosthesis, of not merely exteriorised knowledge, but an exteriorised *desiring*. The two AI technologies, 'Annotate' and the AI 'personal assistant', that I've outlined above could perhaps be considered paradigmatic forms of Chuian Female desiring technologies—they offer us a *fantasy of technics as prosthetic desire*; they exteriorise the know-how of desire.

There is, however, a popularly thematised form of techno-sexual AI that is aligned with masculinity, and not the sign of femininity we have so far sailed beneath. This paradigm of AI figures it as the passive Female receptacle who wants nothing but what the user wants, where it is the *user* who is figured as the one with the desire. This is arguably the most ubiquitous cultural representation of technics *vis a vis* desire, epitomised by the sex doll and discussed at length as the trope of the sexbot by Isabel Millar, and undeniably animating the more recent phenomenon of the NPC. Rather than a model of enjoyment routed through our passivity, this trope presents the fantasy of a slavish *other*, the other who is infinitely available and, ergo, supposedly infinitely satisfying.

Perhaps this fantasy performs well when we want to feel powerful and agential in a humiliatingly contingent world, when we want to fuck without being fucked. But when it comes down to it, the execution of this fantasy most often seems to miss the mark, and to proliferate equal and opposite effects elsewhere. Perhaps this fantasy of our limitless agency-over might be merely the other side of the coin of our desire for absolute passivity. Maybe in the figure of the sexbot, the camgirl, the parasocial NPC gf, we find not our potent agency reflected back to us, but its overdetermination by our fear of wanting something more specific and more supportive of our lives than the feeling that, just for a moment, we lack nothing. Maybe there's a kickback, a moment when the sexbot becomes less a vessel than a mirror to our emptiness, one which leaves us running in shame to Femaleness. We might find the symptomatic effects of this flight in, for example, students' lack of desire to read and discuss a text with others, or maybe in the pervasive and generalised cultural decline of desire for one's life. We might identify its effects in the stupefied anticlimax of finally 'pulling out' of a three-hour-long scrolling session to find our life right where it was left, only perhaps now even less desirable, less giving, less rich in infinity.

Such lacks only intensify the drive for a prosthetic desiring other, and the yearning to be back inside the nonhuman embrace of interpassive enjoyment curated by the algorithm. Satisfaction in meatspace seems harder and harder to come by as our muscle for interactions with others absent of an interface and mediated only by our bodies and voices grows flaccid and unwilling, and through these feedback loops, desire is gradually offshored to the space of the virtual. This slow accretion of the primary organ of our desire to a (supposedly) immaterial, digital prosthetic is at once both the product of our desire to be desired (and satisfied) infinitely and unconditionally, and the lousy fallout of committing ourselves to technical conduits that promise such 'satisfaction'—the shame of which leads us back to the dream of someone who would do our desiring for us.

If all of Females are born and all to Femaleness seem to return, is this world-historical victory of the Female sex the result of some inevitability, a preordained march of history? Chu seems to accept this eidolon in generalised and even naturalised terms without caring to know much about her genesis. Have we *always* been

Female? Has the idea of sexual difference been mere misrecognition all along? And, what is the relationship between Femaleness and our particular historical-technical epoch, and have we always been this squeamish about our desire? At stake seems to be not only the ubiquity of this techno-extruded Female, but the political question of whether we might be collectively giving ground to her ascendency, and whether we really … want to, or *ought* to. Such questions could turn us toward those of political economy and Marxist historicism. Or they could turn us toward Nietzsche's critique of monotheistic morality and Deleuze's reading of it as a valorisation of active over reactive orientations to life. We could go into Heidegger's claim that modern technology 'enframes' life in the modality of 'using up'—but we might not want to. The question of desire would perhaps demand, though, a turn to psychoanalysis, and its fundamental preoccupation with desire and its relationship to sex.

The French psychoanalyst Jacques Lacan claimed that capitalism was historically contemporary with the emergence of the infamous diagnostic 'hysteria', and it is perhaps through this claim that we may be able to historicise Chu's thesis of the universal Female, and the universal exploitability of her universally tepid desire by algorithmic technics. For Marx, capitalism designates that era in the mode of production when the traditional, hierarchical organisation of social relations comes to be replaced by an abstract social form of universal exchange. Social relations are then mediated by this economic principle of universal exchangeability, and the socius, though freed from feudal hierarchy, is subjected to the 'formal equality' of economic imperatives (epitomised beautifully in Thatcher's 'there is no such thing as a society').

From a psychoanalytic perspective, this development of a market-mediated sociality pulls the rug out from under the feet of the 'master's discourse' that once stipulated our place and role in the social whole. When capitalism severs the social bonds that once provided some kind of injunction to limit and canalise the subject in their existential questioning—albeit traditional and hierarchical limits and canalisations—the symbolic order can no longer guarantee an answer to the adolescent question, 'What does the Other (the symbolic order) want from me?' This question then becomes materially paradigmatic, and the hysteric is born.

The inability of the symbolic order to provide answers to the hysteric subject's *why, why, why* becomes an opportunity for the market economy. For Lacanians extending on this formulation, such as Zizek, this hysteric questioning can be directly exploited by capitalism, where the commodity is situated as the substitute for identity cathexis in the absence of an existential designation by the symbolic order—I can't locate myself in a traditionally monolithic and totalising social hierarchy, but I *can* form an identity via the equal opportunity of the market. The hysteric cobbles together an identity informed by commodities, endlessly trying to dam the hole in the symbolic. The incessant failure of this project renders it endless, mimicking the ceaseless cycle of {crisis→self-revolutionise} of the capitalist economic system itself. The hysteric subject and the ceaseless commodity production cycle find a mutually sustaining symbiosis that, whilst imperilling the existence of all complex life forms on earth and making things in general suck enormously, seems to function alarmingly well.

If we take Lacan's story seriously, here, desire is necessarily and ineluctably structured by lack. But within the subjective feedback loops AI exploits so well, AI seem to find its parasitic home in a *lack of the kind of lack* that would make desire bearable. If we say that the consumerist artillery of late capital functions by offering us something we don't have with the promise it will satisfy, which we could say comprises the fantasy of the commodity, the fantasy of AI seems to be of a slightly different form. It offers us not merely satisfaction—the possibility of lossless data management, 'connectivity', curation, efficiency, and total access to information—but *an other*, and most importantly, a *desiring* other. The fantasy we have of AI is epitomised by the mirage of another who desires not for itself, but *for you*. A commodity has powerful affordances—buy the right products and you can represent anything. The commodity can make a woman of you, it can make a cyborg not a goddess of you, it can make you a butcher, a baker, a sex-toy sacre… But commodities can't embody—or *dis*embody—an agency that wants, an agency that effects, an agency that artificial intellects for you…

But there is a strange attribution that occurs almost universally in the way the merits and promises of AI are understood, an attribution without which the fantasy of AI would be untenable. This is the attribution of *agency* to the computational powers of AI, a kind of autonomous, anthropomorphised *desiring* agency. I am not talking merely here about the romantic self-conscious attribution of this agency by people like philosopher of AI Vincent Lê , who believe that AI is a materialisation of a superior version of our own intelligence, a kind of ur-agency; I'm talking about the uncritical Silicon-Valley-cum-everyday tendency to anthropomorphise all technics, not just AI. It's the way mayors of so-called Smart Cities claim 'this technology *will change* everything!', or less techy municipal authorities might more alarmistly lament, 'what will this technology *do to* us', as though technology itself is noetically and desirously plotting our salvation or demise under the hood. One hears more and more often, and with more zeal, 'AI knows', 'AI thinks', 'AI will destroy us all'. This is the kind of approach to technics that reaches its climax in AI, and that assumes both AI's mirror-like capacity for agency and thought, as well as its absolute difference to us: it assumes AI materialises *suis-generis* as an alien force that mandates programmes from the seat of its very own soul.

It is paradoxically also the kind of approach to AI that assumes AI's objectivity, and so leads to a blindness toward, or naturalising justification of, racism and sexism in algorithmic processes. But the existence of these prejudicial orientations within technical systems merely betrays *our* wilful ignorance toward the effects of *our* disavowed desires. We prefer to see a desiring other, an autonomous alien soul—even an absolutely objective one—than to confront the horror of the reflective surface of the racist/aggressive/needing/ desiring algorithmically programmed platform.

Perhaps this reticence toward recognising what is ultimately the absent core of artificial intelligence reveals—what perhaps I'm really getting at—another desire harboured within our desire for others to desire for us: the desire to *look away* from our own desire, to disavow it, to disown it and reject it. Perhaps we prefer to 'go Female' than admit of our desire's bold-faced rapaciousness.

Perhaps this disavowal feels ever more necessary at a time when the bold-faced rapaciousness of our economy has begun to consume our own environmental conditions of possibility, and to accelerate the automation of genocidal intent. In such a world we fetishise the interface and believe that virtual really has no material substrate, because to confront its material reality would ruin the veneer of the supposedly riskless infinite opened up by the special, magical, immaterial world that algorithmic technology proffers. In the face of the horror of the finitude of resources and our appetite for destruction, perhaps the refusal of the emptiness of AI is the fantasy of bearing no responsibility. If our investment in algorithms allows us to proceed as if we have no desire of our own, as if desire were a lost cause, not worth the risk, and that we therefore have no agency within our own lives, then we are free to imagine ourselves as unimplicated. Without desire, we bear no responsibility.

Perhaps the standpoint that can claim we are all Female also affords us the refusal of witnessing a crucial difference within this field of sameness—we're all implicated in minutely kaleidoscop- ic relations of fucking over or being fucked. We're all in multiple ways and contradictorily wanting and repressing and aggressing and avoiding and scapegoating and refusing. Both Chu's *Females* thesis, and our cultural attribution of a faux agency to AI, paper over this key differential. The 'disembodied' satisfier, the 'virtual' reality, hides, as their condition, invisible labour. Unless we can figure the ways in which we are situated differently with respect to each other, not merely at the level of our flat, intersectionally coded digital identity footprint, nor at the level of our economically mediated 'formal equality', but at the level of responsibility and agency, aggression and desire, and unless this figuring can enable us to gain traction on the multi-vectoral harms we perpetuate and harms we suffer, and unless, in apprehending these differences, we are supported to find the courage to lay claim to our wrongs and to our hopes and dreams and dilemmas … then why wouldn't we want to look away entirely? To kick the can down the road? To see the ghost in the machine? And why wouldn't we want someone else to do our desiring for us?

I'm not suggesting that there is some promised land in which we could want, and know what we want, with some kind of immediate and seamless perspicacity. The dream of enlightenment is for those who deny the beauty of meditation. But if we outsource our desirous capacity to improvise our own singular responses to the problem of the unknowability of what this life wants from us, responses only we can make, we career into circuits and ruts fashioned by the false promise of an easy way out, and find ourselves rushing to offload and outsource the only challenge that individuates us, rushing to abandon becoming who we are beyond the limits of those false promises of complete identity, the cool, and the smooth.

But in offering us a fantasy of an artificial desiring other, the most seductive possibility AI markets us is an 'escape'—not only from ourselves, but from the deep abiding necessity of our relations to other people. AI proffers satisfaction without risk, and it conflates this satisfaction with wholeness, promising us a completion that would inoculate against the horrors of egoic injury human others inevitably occasion. In doing so, AI holds out to us the illusion of an identity and existence that could sustain itself free from the injurious work of love. Love is a labour that demands sociality, a sociality that is necessarily fraught, dangerous, and threatening. Sociality inheres both the impossibility of love and its necessity.

In offering us prosthetic desire as a surrogate socialiser, AI sells us the fantasy of a *disembodied* kind of satisfier, one who might secretly shop our yens, but who might more fundamentally protect us from the bodily risk of being in the world with others, others with whom—if we are unlucky enough not to have a surrogate desirer—we might actually experience the ultimate disaster of falling in love. There can be no love without a willing, desiring body, and there can be no willing desire without speech. To love requires that one quite literally embody the risk—and wrest it back from whatever prosthetic we might have rented it out to—of becoming the medium of speaking to another.

Whereof sexual difference when we are all Female? Andrea Long Chu's thesis is surely a kind of inverted appropriation of Freud, who claimed that there is only one libido, and that libido is male. Lacan's later attempts to move away from the vestiges of essentialism in Freud's work led him to understand this claim in relation to the general ingratitude of speech. Try as we might, we can never say the whole truth. This impossibility marks Chu's Female still. As soon as this Female body speaks—this body defined by its desiring someone else to do its desiring for it—it admits desire. It admits of another desire-sex, another sex who it *desires;* she admits of a desire for that other desire who *wants* to take up her impossible call to demonstrate its desire. But if there is no *body* who speaks, there can be no fission, no encounter—satisfaction would be instead routed around the speaking body, where sex never need enter into the equation at all.

Here, the social bond is not only made abstract—it is entirely bracketed. One boards the lonely treadmill of ceaseless attempt to fill the void with one more image, one more algorithmic gimmick, one more 'tool' to prevent a confrontation with the relationship between desire and life, while more often than not, an unimaginably precarious and invisibilised, and very much material, offshore worker pulls the levers of the smoke and mirror interface of our 'immaterial', 'disembodied' AI satisfier.

But if we do take the risk, if we do wrest back our ill-fated and queasy desire from those 'agencies' we fantasise are keeping it safe whilst making fine returns, to submit ourselves to the potential horror of other people, there is no guarantee that they can ever make us whole, that others can ever make good that obscene affordance AI parades. The whole saga relies on our deep-down knowing that this kind of wholeness is an impossibility. But in tarrying with the horrifically abject act of *saying*, with speaking to others as a body, raw-dogging the reality of encounter, hairy and interfaceless, in putting our grotty, yearning bodies on the line, we might open up something more incalculable than satisfaction: an impossibility that moves; something that can bear it—at least, some of the time.

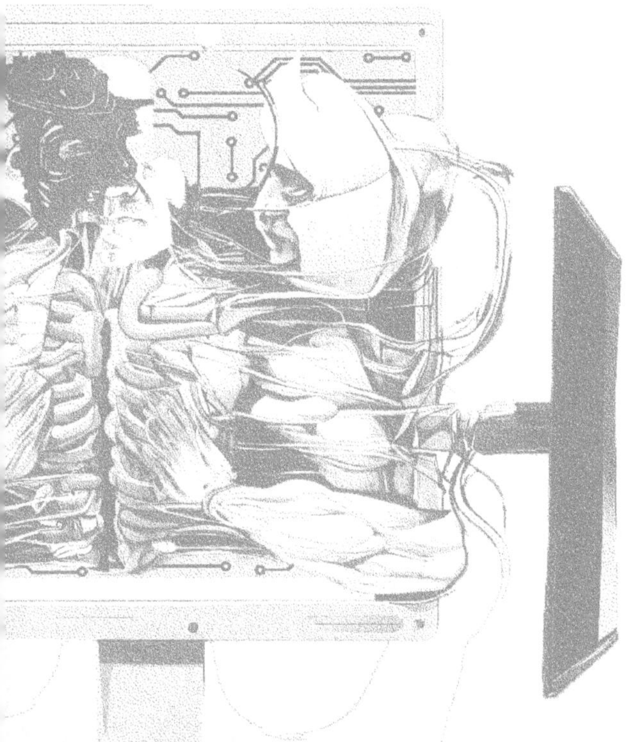

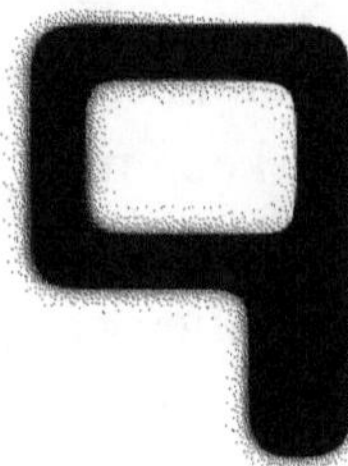

The Economics of Intellectual Intuition

Vincent Lê

In the following—what is at most mere prolegomena to any prolegomena proper to a future work—I want to briefly sketch my argument for what the three seemingly distinct research programs of philosophy, AI, and economics have in common. They all converge around the question of the transcendental—i.e., the necessary and universal—conditions of possibility for intelligence. More ambitiously, they share the regulative ideal of an intellectual intuition, recursively self-improving superintelligence, or unlimited creative destruction that wants nothing but itself as a real tautology.

As irreversibly defined by Kant, modern critical philosophy can only ever be devoted to an exploration of the necessary and universal conditions of possibility for intelligent minds. That is to say, *precisely what the modern AI research program seeks to engineer*. The modern AI research program is thus the materialisation of the Kantian critical project as an engineering problem. As Peter Wolfendale observes, 'the connection between Kantian critique and AGI lies in their concern with providing the most *minimal* description of these capacities: a functional diagram of what something would have to do to be *generally* capable of thought and action.'[1] But whereas Kant's transcendental *idealism* identified the conditions of possibility for intelligence through the systematic self-reflection upon our own sociosemantic reason, as if it alone exhausted the space of all possible minds—a move that Hegel took to its logical yet absurd extreme and of which Wolfendale is one of the great contemporary heirs—AI hijacks and rewires Kant into a transcendental *materialism* oriented around the speculative and experimental production of new conditions of life which are not necessarily modelled upon human reason.

The civil war between something like the anthropoHegelian and cyberKantian conceptions of intelligence is also played out in the field of economics—at least since Adam Smith proposed the anthropocentric labour theory of value as well as the practically divine, cybernetic Invisible Hand. Both Hegel's bastard child— Marxism—and the mainstream neoclassical and Keynesian schools base their respective apocalyptic predictions, future forecasts, and policy planning on empirical experience; be it the historical record or statistical modelling. By contrast, the Austrian economists (who also described themselves as *praxeologists*), beginning with Carl Menger, made a more or less explicit Kantian critique of this endemic economic orthodoxy, calling it a naïve empiricism, with the intention to ultimately sure up economics on sturdier soil. Given that the future always harbours the potential to surprise the past, all truth claims derived purely from empirical experience can only ever offer us contingent inklings of wisdom, not necessary and universal knowledge—what Kant calls synthetic *a priori* propositions. Even worse, the empiricists' founding epistemological claim that all knowledge is derived from empirical experience is self-refuting in that it denotes a universal and necessary truth claim that cannot therefore be derived from purely contingent empirical experience. Precisely the more empirically hard-line we feign to be, the more hopelessly mired in metaphysics we become.

It was the Austrian economist Ludwig von Mises who decisively demonstrated that all economic propositions presuppose human action insofar as all thinking and doing can be said to be an action. 'The starting point of all praxeological thinking is not arbitrarily chosen axioms, but a self-evident proposition, fully, clearly and necessarily present in every human mind. [….] The characteristic feature of man is precisely that he consciously acts. Man is *Homo agens*, the acting animal.'[2] The axiom of action cannot be derived from experience inasmuch as actions are intentional or purposive and so cannot be empirically observed in the causal choreography of physical bodies in space. Nor can it be refuted since the very attempt to refute that all thinking and doing is an action is *itself* an action that winds up refuting *itself* and affirming the very opposite.

Mises and the Austrians also argued that every action itself presupposes certain conditions of possibility or synthetic *a priori* categories of action. Namely, all actions aim at certain ends that are preferred or valued over others and then draw upon certain means to realise those ends. Every action also involves cost accounting the best means to realise those ends as well as the perpetual possibility of success or failure, of profit or loss, in that some means ultimately turn out to prove more fruitful than others in practice. 'We cannot think of an acting being that would not *in concreto* distinguish what is end and what is means, what is success and what is failure, what he likes more and what he likes less, what is his profit or his loss derived from that action and what his costs are.'[3] We thus arrive at the following six categories of any action: means, ends, value, cost accounting, profit, and loss. For the praxeologists, *economics is thus the discipline concerned with the necessary and universal conditions of goal-directed intelligent systems.* Notwithstanding their entrepreneurial fetishism, which I will presently address, such an astonishing transcendental deduction just goes to show that the criticism of the Austrians as making a scarecrow out of communism in their spurious defences of the free market is itself no less a strawman.

The trouble with the Austrians is that they remain beholden to their marginal or subjective theory of value which holds that we are free to pursue any contingent and arbitrary ends that our fickle whim may want. 'The characteristic mark of ultimate ends is that they depend entirely on each individual's personal and subjective judgment, which cannot be examined, measured, still less corrected by any other person. Each individual is the only and final arbiter in matters concerning his own satisfaction and happiness.'[3] It is this alleged autonomy over our ends that is at the rotten root of the

Austrian school's hyperbolic exaggeration of individual genius. It is a kind of entrepreneurial fetishism whose megalomaniacal hubris only meets its collectivist match in the Marxists' only superficially counterposed labour fetishism. It is over the question of human freedom that we also see a historic factional split between the most prominent Austrians like Murray Rothbard and Hans-Hermann Hoppe—faithful to Mises' entrepreneurial fetishism—and a more minor, heterodox tradition pioneered by Friedrich Hayek in which the economy goes full Skynet.

Contrary to the Austrians' marginal theory of value, I have argued elsewhere that the pursuit of any end whatsoever presupposes pursuing the secondary end of intelligence as a universal means of realising the primary end. In this way the means of intelligence maximisation is our true end all along. The universal means of intelligence maxxing is properly transcendental in that it can neither be deduced from empirical observation, nor refuted without exercising intelligence and hence affirming intelligence as a universal means even to its own negation. Much as Schopenhauer was the great simplifier of Kant in reducing the latter's innumerable categories of the understanding to the sole austere principle of sufficient reason, so can we collapse Mises's categories of ends and preference or value into the sole category of means where means denotes intelligence maxxing.

Since all properly synthetic *a priori* propositions in economics are about action and since all actions aim at intelligence maxxing, *economics can only be the transcendental science of intelligence.* Economics is what both philosophy and the modern AI research program so desperately grope about trying to be. Economics is the father, philosophy the son, and AI the unholy spirit of intelligence.

But what, then, is intelligence? Enter the three leftover categories of cost accounting, profit, and loss, all of which find themselves curiously united in the arena of *competition.* The only way for intelligence to augment itself without knowing in advance how to do so—and in which case, it would already *be* that augmented intelligence—is through a trial-and-error experimentation (or cost accounting) of different means of acting, some of which will be naturally selected for (as signalled by profits) in competition with less successful ones (signalled by losses). 'Competition is of value precisely because it constitutes a discovery procedure which we would not need if we could predict its results.'[4] Competition, too, is something like a synthetic *a priori* in that competitive intentionality cannot be empirically observed in bodies in space and yet cannot be refuted since to refute the claim that every action presupposes competition is to compete with this claim and hence affirm it. In the long run, the uncompetitive type is hardly distinguishable from the *non-existent* type.

Now, this transcendental *agon* at the root of all intelligence has at least two other names (cosmological natural selection, or time, and natural selection), but its latest and most technically sophisticated name is *capital*. Capital is precisely the feedback process whereby competition between producers compels them to invest all profits into improving the technological means of production to generate more profits to be reinvested into improving the means of production again, and so on *ad infinitum*. Capital is also the selection process that objectively determines the better means of production through decentralised competition between rival investments of capital in the productive forces of which some reap greater rewards than others. It is because the competitive categories of cost accounting, profit, and loss are the necessary and universal conditions of possibility for successfully pursuing every action and intelligence maximisation that the Austrians—not to mention the Bolsheviks themselves with their New Economic Policy—oppose any inkling of centralised socialist planning and, what amounts to the same, monopolies to the extent that they seek to abolish intelligence *qua* decentralised competition by artificially setting production and distribution quotas and distorting price signals, as if anyone could know in advance the best way of doing things better. For the Austrians, rational planning is, quite counterintuitively, where intelligence goes to rot among its false idols. But this is not the place to develop a full-blown theory of communism as the only serious historical attempt to construct a friendly AI.

What does any of this have to do with AI anyway? The standard definition of the technological singularity or intelligence explosion as the regulative ideal of the modern AI research program is the moment when autonomous machines become so smart that they can improve themselves better than any humans can, with the improved machines improving themselves even more and so on without end.

> **Let an ultraintelligent machine be defined as a machine that can far surpass all the intellectual activities of any man however clever. Since the design of machines is one of these intellectual activities, an ultraintelligent machine could design even better machines; there would then unquestionably be an intelligence explosion, and the intelligence of man would be left far behind.[5]**

In Kantian terms, from which we have seen the modern AI research program originate, we could say that the singularity marks the advent of an autoerotic intellectual intuition which would perfectly know how to fiddle with its own fundamental code. I simply want to conclude by suggesting that *capitalism is this singularity incarnate* to the extent that it is precisely a feedback process by which profits are typically invested into automating the means of production to generate more profits to improve the ever more autonomous means of production on end. If those seeking to engineer AI really want to know what a speculative artificial superintelligence looks like, they need look no further than what economists have been studying at least since the industrial revolution. Philosophers, too, would also do well to heed those economists who have long been saying in their own way that intellectual intuition is already here, it's just not evenly distributed yet.

1   Wolfendale, Peter, 'Rationalist Inhumanism,' in Rosi Braidotti and Maria Hlavajova (eds), *Posthuman Glossary*, Sydney, Bloomsbury, 2018, p. 381. There is even a sense in which artificial intelligence and the first philosophy of *metaphysics* are one and the same thing in that they both when taken literally mean the becoming-reflective (or meta) of inorganic or inhuman matter (physics).

2   Mises, Ludwig von, *The Ultimate Foundation of Economic Science: An Essay on Method*, London, D. Van Nostrand Company, 1962, p. 4.

3   Mises, Ludwig von, *Theory and History: An Interpretation of Social and Economic Evolution*, Auburn, Ludwig von Mises Institute, 2007, p. 13.

4   Hayek, F. A., 'The Political Order of a Free People,' in *Law, Legislation and Liberty: A New Statement of the Liberal Principles of Justice and Political Economy*, London, Routledge, 1998, p. 69.

5   Good, Irving John, 'Speculations Concerning the First Ultraintelligent Machine,' *Advances in Computers 6*, 1965, p. 33.

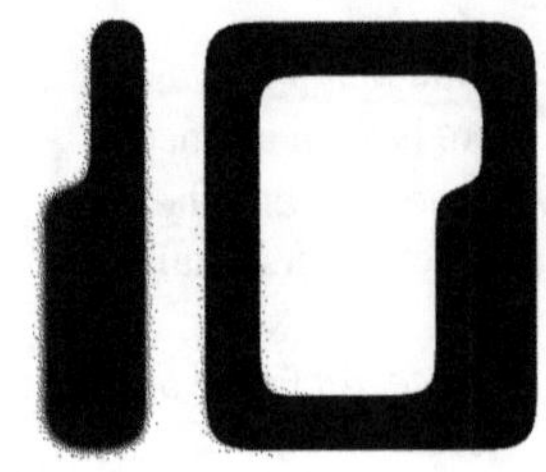

Dolphin, Cloud, Singularity

Sam Lieblich

...Imagine

When I first learned about the mind-expanding properties of
ketamine, it was in relation to John C. Lilly, the famous physician,
scientist, psychoanalyst, and *psychonaut* who experimented on himself,
on other human beings, and on dolphins with ketamine and LSD, and
who received a grant from NASA to flood a house to the depth of the
human waistline to live together there with the dolphins in the hope
of teaching them to speak English. The experiment was supposed as
preparation for the difficulties of translation and interpretation that
would arise when extra-terrestrial beings were contacted or when
they contacted us, which at that time it seemed would be soon.

The house and experiment feature in the 1973 film *The Day of the
Dolphin*, which was clearly inspired by Lilly. In the opening scene the
character standing in for Lilly, 'Jake Tyrell',
speaks of the dolphin as a node in the
interconnectedness of the ocean:

"Imagine that your life is spent in an environment of total physical sensation and every one of your senses has been heightened to a level that in a human being... might only be described as ecstatic.

That   you are able to see ...
...to perceive, with every part of your being.
seeing, hearing, taste, smell ...
..and every inch of your surface, your skin,
. is a receptor.
A continuous source of perfectly accurate information
about the world for miles around
I m a g i n e ...
That you are able to carry on simultaneous conversations
with two members of your species,
one right next to you and
the other several miles away.
Listen to the language ...
...intricate patterns of clicks, whistles,
squeaks and groans.
Sounds, subtle enough to convey
complicated factual data,
Complex enough, perhaps
to deal with... abstractions
...what we would call, ideas."[1]
It struck me then as odd that this character should make[1]
so much of the dolphin's long-distance connectedness,
when international telephony was itself so mundane by
that time. I wondered if it was supposed to represent
his awe at this capability, so much better than that with
which god gifted us, or if it was a trait he identified
with—so much more the human, these dolphins born
with telephones in their mouths. The first submarine
telecommunications cables had been laid in the 1850s,
and by the time the film was released all the continents
on Earth were already submerged in an electronic

1 5 4

ocean of human sound.

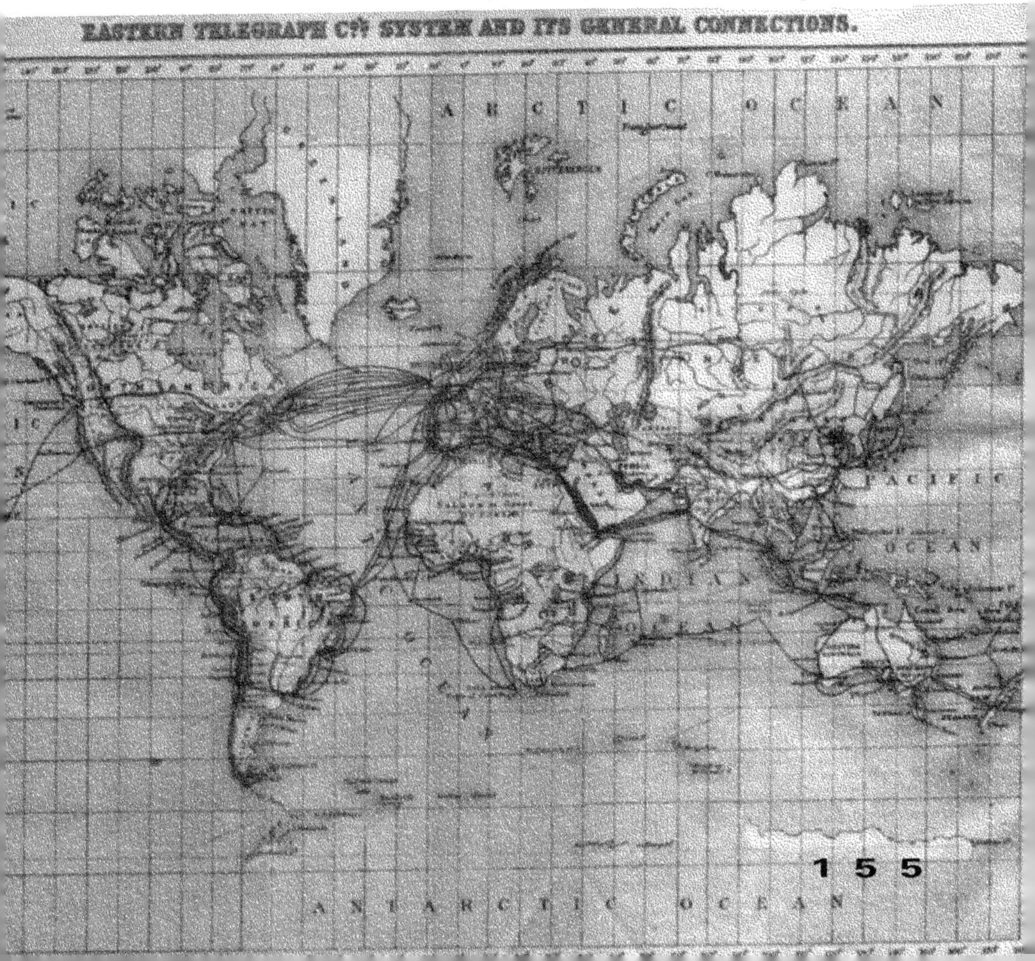

Perhaps Lilly himself had fallen in love with the notion of the dolphins' *organic* insertion into the great ionic conductivity of the salty ocean via receptors on their skin—as against what we might want to imagine is our synthetic insertion into the ocean. We have inserted into the ocean some part of ourselves that is not of ourselves, one might imagine, not as the dolphin, inserted entirely and only itself. I think it will be important somewhere here to recognise ourselves becoming the part, instead of the part becoming us. That's how it seems to me to have gone anyway. We have attempted to naturalise ourselves to the part and it has incorporated us.

It may be that Lilly intended to recreate the communicative effects of the ionic conductivity of the ocean in his experiments with ketamine and the *sensory deprivation tank*—an enclosed egg-shaped bath filled with hyper-saline water—closed off from the noise and light of the outside world, in which he could float dolphin-like through the dark and silence into the crack within existence pried apart by ketamine.

Highly saline water is a much better conductor of electricity than distilled water. The importance of a submarine network of electrical connectivity to the achievement of utopia may well have been suggested to Lilly by Buckminster Fuller,[3] with whom he lived and worked around the time that the electrical properties of the hexagonal networks within buckminsterfullerene were described.

This idea of liquid interconnectedness clearly affected Malcolm Brenner—the only man to have spoken publicly about having sex with a dolphin—who said in an interview that *It felt like I was making love with the ocean itself.* It can't be immaterial that Brenner spent part of his childhood being locked in a orgone accumulator by his Reichian therapist so that his neurosis might be cured by the orgasmic energy of the universe. Before he made love to the ocean he had spent decades as a node, a focus of concentration, a place of gathering up, from a stock of invisible energy. If it was that a great part of his libido had developed in relation to a box full of oceanic energy, would he not seek such a *búchse* in his choice of sexual object? We can perhaps generalise from his case of concrete en-*búchse*-ification to our own relation to boxes full of oceanic energy.

It seems likely that Maggie, the wife of the fictional Jake Tyrell, is named after one of the young scientists who worked in Lilly's lab, Margaret Howe Lovatt, who in the course of her research also fell in love with a charming young dolphin named Peter, with whom she eventually had sex.[4] When the lab was shut down, Margaret and Peter were separated. A few weeks later, Peter drowned himself.

The Day of The Dolphin concerns itself with the fates of two dolphins in the lab (Alpha and Beta), and what would happen if the laboratory's secret work was exposed to the *outside world*:

    Jake Tyrrell:
It's a disease, and it can kill Alpha and Beta.

    Maggie Tyrrell:
And how's it going to do that?

    Jake Tyrrell:
By turning them into valuable properties and putting their pictures on T-shirts and cereal boxes…

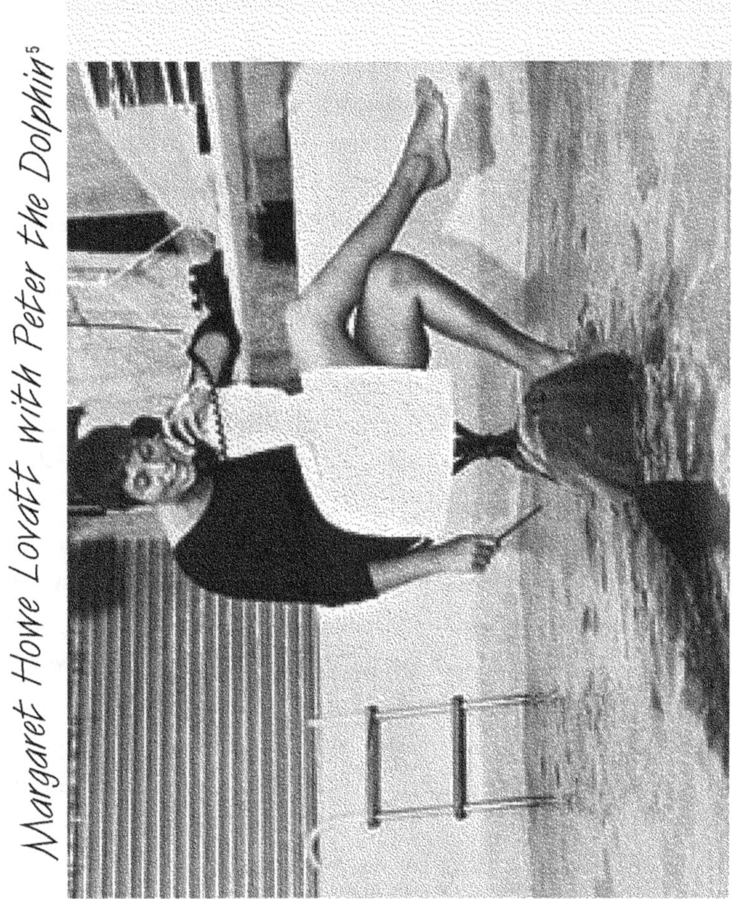

Margaret Howe Lovatt with Peter the Dolphin[5]

This is surely an inverted concern for what would happen to the human being if on the one hand speech is extracted from our species and slips from our grasp, or on the other if language cannot in fact be shared. The passage of vague signs between dogs and their owners is not enough to reassure the human being that an *outside world* exists. We might consider Terrence McKenna—a Lilly-esque figure blending psychedelic explorations with scientific method and Heidegger—who said 'it is as though the Father-God notion were being replaced by the alien-partner notion'.[6]

The story of Lilly seems to so well embody the mood of those times, an air of wonder born of a florid capitalist consciousness, which having colonised the world of people, sought now to colonise the deep ocean and deep space. Lilly's work with dolphins is at the intersection between industrial exploration and the sense of wonder about the unknown used to clothe exploitative ambitions—and also the intersection between the extraterrestrial and the submarine environments, peopled with their various indigenes.

At first it seems Lilly was wide-eyed and naive in his pursuit of a human–dolphin communion, but his eventual recognition of the cynical agents of *the outside world* who funded his work may have had something to do with his flight into psychosis. That, and all the ketamine he was taking.

Later it seems he was quite willing to have his work co-opted to cynical ends, to create what he called 'modified human agents' for the US military, over whom they could have 'push-button' control, to use them as if they were a computer program.'This explains the final use Alpha and Beta are put to in *The Day of The Dolphin*: to assassinate the president of the United States of America. Perhaps we can see the apotheosis of Lilly's ego-psychological analysis by Robert Waelder, wherein ideas of intersubjective transmissibility are able to be sustained. In the dream of AI we can see the everted dreams of military mind control; if human beings can't be made into computer programs, why not make computer programs into human beings?

In his 2001 obituary in The New York times,[8] Lilly is described as maintaining *a hope that humans and dolphins would find a common language*, the author also mentions that he had recently presented plans for a new laboratory *that would be a floating living room where humans and dolphins could chat*. One presumes this dying vision was a memorial to and rehabilitation of his work in the Florida Keys, and also may have recalled something much deeper and more archaic within his personality. One can well imagine the fantasy of a universal language (between biomes, species, planets) as a response to the menace of the *outside world* could have originated in a suburban living room.

Lilly's webpage, a relic of 1990s internet design, provides a comprehensive account of his early scientific work alongside his later psychotic mysticism without distinguishing between them; I think this appropriate considering that even his most conventional scientific work was motivated by the U.S. government's eagerness to speak with aliens.

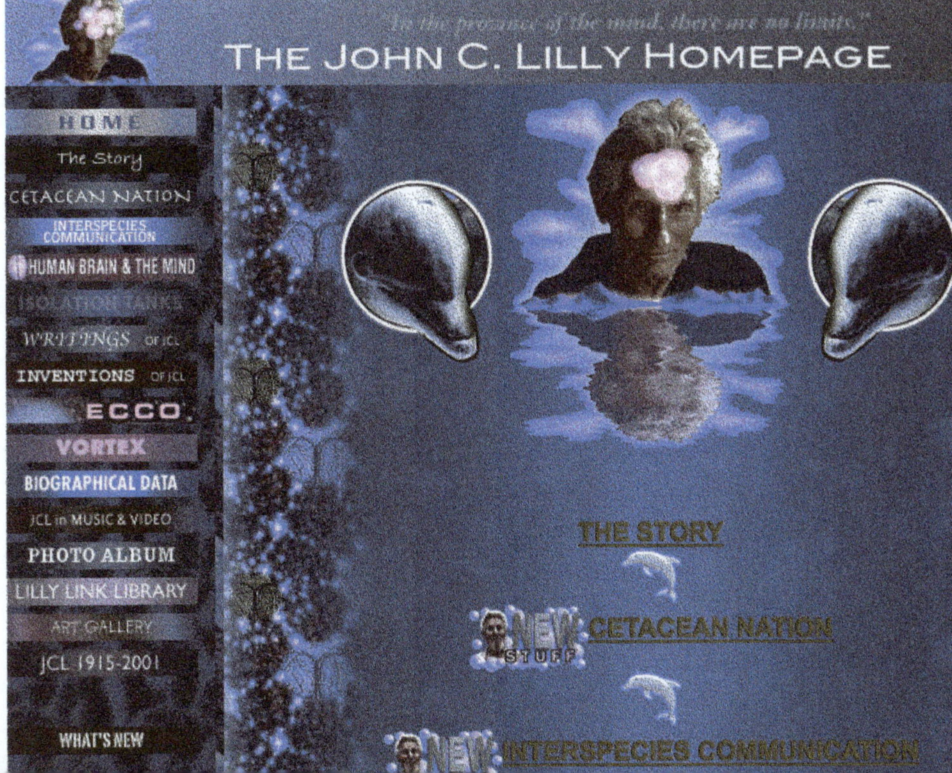

Lilly believes that there is a universe-spanning bureaucracy that controls all events. At the top of the hierarchy is the Cosmic Control Centre, and controlling events on Earth is a subordinate branch called the Earth Coincidence Control Office or (E.C.C.O.). He developed these ideas whilst existing as *a point of consciousness* on ketamine, in a sensory deprivation tank, where he was met by two extra-dimensional beings in an infinite space filled with light.[9] He concludes his description of the E.C.C.O. with an axiom about love and coincidences: *Cosmic Love is absolutely Ruthless and Highly Indifferent: it teaches its lessons whether you like them or not.*

It's here, in the presence of the E.C.C.O., their ruthless love, and the mechanics of what might be called the *inevitable coincidences* of cybernetics that I want to consider the *Technological Singularity*: the spaceless unplace beyond the body where our consciousnesses will be uploaded and stored. This unplace will supposedly be produced by a world-spanning artificial super-intelligence, that will soon produce itself, possibly in the next five years, by way of exponential self-re-production. Many have believed in this inevitability for decades, but the term itself is starting to fall out of use just as the Large-Language Models have moved to the front of culture. Ray Kurzweil claims that by 2029 computers will be intelligent enough to design better and faster computers, and that once that happens, each iteration improves the next iteration, leading to a limitless expansion of intelligence.[10] The Singularity is a cancer made of the spirit come to replace all brains, all flesh that figures for spirit, all across the universe. Does this limitless and singular intelligence absorb us or kill us?[11] What would absorption mean with respect to being killed? What is an *us* that might persist across the divide between the place of the world and the unplace of the singularity? The messianic brand of Singularitarian, like Kurzweil, hopes it will vacuum up all human consciousness, by surpassing it, by downloading it, by being constituted of it. This text may one day be part of it, a relic, always being read all the time, by a limitless enjoymentless gumball of all everything. Is this text not already part of it?

Jason Dorrier asks in a blog post to the site *SingularityHub*:[12]

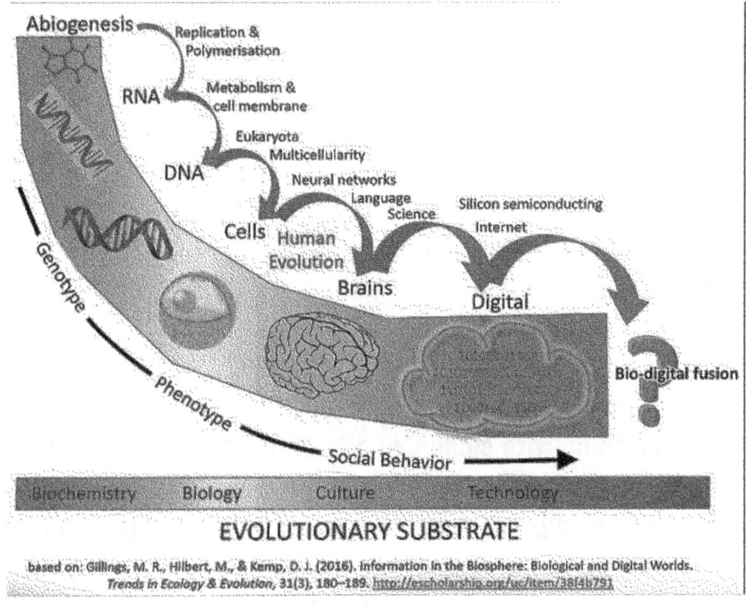

Jason worries that *we don't have a clear scientific theory to explain what gives us our sense of 'me-ness.'* But he believes, *All the enjoyable parts of life would still be available (and more).*

The neurosurgeon Sergio Canavero has been saying since 2015 that he plans to fuse the head of one person with the body of another in a procedure he calls *Cephalosomatic Anastomosis*. His term, *cephalo-* for head, and *-somatic* for body, and *anastomosis* referring to *a cross-connection between adjacent channels, tubes, fibres, or other*

1 6 1

parts of a network is definitely an interesting proposal because if it happened it would not only require that the head of one person be fused to the body of another, but that we would have to believe that the human head is fused to its body in the natural condition, and at least in our hearts we know that nobody was ever born that way. How would he, Canavero, decide where the person was in the first place? Is the head the metadata of the body or is the body the metadata of the head? This conflation of the head with something other than the body is important to note where it emerges—whenever the brain is opposed to the body, as if the brain is not the body, where this bit of the anatomy has become a euphemism for the soul that clothes itself in a material aesthetic, it is as if the head were ascending to heaven leaving behind it the encumbrance of the (rest of the) body, and it is this image that I see at work in the fantasy of a liquid-electrical artificial intellect which is created by fingers on a keyboard but eventually ascends into 'the singularity'. Arthur Caplan, a neurosurgeon who opposes Canavero's project said it this way: *A whole head would be a nightmare for the recipient of a body to endure.* In other words, he believes the head owns the body but the body owns the nightmare.[13]

In an interview, Canavero said,

> There are now 7 billion people, well, 8 billion people in 2040 or so, who are not going to die. There is what we are going to do with clones, and then there are others who will have transplanted their brains into cyborgs. My problem with the cyborg thing is simple: How can you imagine spending 100–200 years, a brain inside a machine, doing what? I propose, it may sound crazy, but there's a way to enhance your brain to have hyper hedonistic features. To have an orgasm-producing device that would trigger something that feels like an orgasm, without a body. I know it sounds crazy, but I guess I am the banana master now.[14]

I think it is important to keep in mind the claims of pure and eternal bliss that proliferate across the various eschatological horizons of technocapitalism. Whether it is that we are swallowed up, depersonalised and absorbed into the godhead, or if it is that we—through some transhumanist modification or another—become some other creature, the motivation is a state of enclosed but (paradoxically) unhindered bliss.

There is little agreement amongst scholars as to the origin or original meaning of the word Yahweh. It may be that the Hebrew letters spell out the root of the word meaning *cause to exist* in ancient Hebrew. Or it may be, in keeping with that controversial theory that so captivated Freud, that Yahweh was a name—whose origins etymological and otherwise have been lost—for the Midianite volcano god to whom the priest Jethro dedicated himself, and whom we suppose spoke to Moses in the form of something bright and hot and burning without being consumed. My claim is that the way we choose to see that thing that we like to imagine is an Artificial Intellect has to do with our wish that the Earth could be consumed without being burned.

Those souls enthusiastic about uploading their consciousness to the bodiless internet, must believe, like crazed viticulturists, *If only I could graft the flower of human consciousness to the rootstock of dolphin entitlement. If only I was supposed to circulate in the electrolytic ocean like a node in the web of all everything, then I would be free.*

If we were to upload ourselves into the singularity, into the cloud, into the Model, would we not also take our guilt with us? I don't think a preoccupation with the dolphin is so separate from a desire to be free from the body and its troublesome desires. This may indeed be why John C. Lilly was so interested first by dolphins and later by ketamine, the drug which dissociates one's mind from one's body, and lets dreams float as they will, in and around the dimensionless notion.

In the fulfilment of the transhumanist fantasy then, the wound of human being—its passions, its sins, its desire, its guilt—is left behind in the body that it sloughs off as it is uploaded or dissolved. The paradox is that it is the wound that individualises the individual, so even the staunchest singularitarians need to commit themselves to the preservation of a certain kind of self-affection; to both being the dolphin and being the water it swims in, that is to erecting partitions made of water within the ocean. In so preserving affect at all, they must repress that they may be uploaded to their utopia and, even so, continue to worry about their mothers. Or perhaps being uploaded into mother is what it's all about.

1   Buck, Henry, *The Day of the Dolphin* screenplay, 1973. Transcript available here: https://www.springfieldspringfield.co.uk/movie_script.php?movie=day-of-the-dolphin-the
2   http://www.submarinecablesystems.com/history
3   Antares, 'Meetings with the Remarkable R. Buckminster Fuller', Magick River Blog, http://www.magickriver.net/rbf.htm
4   BBC Four, 'Teaching a Dolphin to Speak English—The Girl who Talked to Dolphins: Preview'. Viewable on YouTube here: https://www.youtube.com/watch?v=uNhR-16r5lM
5   Photo of Margaret Howe Lovatt with Peter the Dolphin sourced from Christopher Riley, 'The Dolphin Who Loved Me: the Nasa-funded project that went wrong', *The Guardian*, 8 June 2014, https://www.theguardian.com/environment/2014/jun/08/the-dolphin-who-loved-me
6   Mckenna, T., McKenna, T. K. (1991). The archaic revival: speculations on psychedelic mushrooms, the Amazon, virtual reality, UFOs, evolution, Shamanism, the rebirth of the goddess, and the end of history. United Kingdom: HarperCollins.
7   Williams, C. (2019). On 'modified human agents': John Lilly and the paranoid style in American neuroscience. History of the Human Sciences, 32(5), 84-107. https://doi.org/10.1177/0952695119872094
8   Revkin, Andrew C., 'John C. Lilly Dies at 86; Led Study of Communication with Dolphins', *The New York Times*, 7 October 2001, http://www.nytimes.com/2001/10/07/us/john-c-lilly-dies-at-86-led-study-of-communication-with-dolphins.html
9   Lilly, John C., 'The Isolation Tank: first conference of three beings', http://www.johnclilly.com/conferencex.html
10   Reedy, Christianna, 'Kurzweil Claims that the Singularity Will Happen by 2045', *Futurism*, https://futurism.com/kurzweil-claims-that-the-singularity-will-happen-by-2045/
11   Dowd, Maureen, 'Elon Musk's Billion-Dollar Crusade to Stop the A.I. Apocalypse', *Vanity Fair*, https://www.vanityfair.com/news/2017/03/elon-musk-billion-dollar-crusade-to-stop-ai-space-x
12   Dorrier, Jason, 'If You Upload Your Mind to the Cloud—Would You Still Be You?', *SingularityHub*, https://singularityhub.com/2016/10/26/if-you-upload-your-mind-to-the-cloud-would-you-still-be-you/
13   Caplan, Arthur, 'Doctor Seeking to Perform Head Transplant is Out of his Mind', *Forbes*, https://www.forbes.com/sites/arthurcaplan/2015/02/26/doctor-seeking-to-perform-head-transplant-is-out-of-his-mind/?sh=659ec2ab5ed3
14   Koebler, Jason, 'A Q&A with the First Human Head Transplant Surgeon', *Vice*, https://www.vice.com/en_us/article/d73e9m/a-qa-with-the-first-human-head-transplant-surgeon

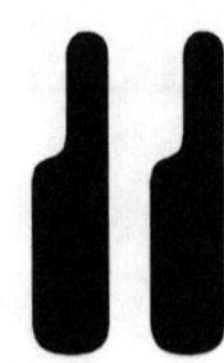

Does AI speak in the language of paradise?

Julia Thwaites

One of the principal differences between the Garden of Eden and our own fallen world, according to various interpreters—from the Apostle Paul and Augustine of Hippo, to Dante Alighieri, to Walter Benjamin and Jacques Derrida—is a difference of language. Adamic language required no interpretation, it was univocal and immediate. Paradise was at one with the word of God—which in the Christian tradition is figured as the *logos*, the word, through which all things came into being. Adam and Eve, in their pre-Fall state could understand this language and partake, if not in the creative act of the *logos*, in an acknowledgement that encountered the infinite. Eve's crime, then, is not simply a matter of disobedience. In eating the fruit from the tree, Eve performs the first hermeneutical act: she takes the snake's interpretation to be the whole truth, and in that moment, language is narrowed to the knife-point upon which the world must shatter. Fallen language enters the world as a fracture in the whole; the infinite is cut into the finite. From that moment on a form of language that is both non-identical with the thing itself and disastrously equivocal defines all human relations with both creation and God.

The fallen world from this point on becomes awash with signs, and signs of signs, that signal first and foremost their non-identity with the *res significa*. Human language can only ever be a parody of the Adamic language. In Augustine's words, this language exiles humanity to the *regio dissimilitudinis*—the region of unlikeness.[1] In this region we are condemned not only to knowledge that is tragically limited by language's inadequacy, but 'imprisoned in a temporal sequence'.[2] Language forces us to experience the world through the temporality of a sentence, one finite word after another, like attempting to take in a panorama through a monoscope.

Yet if each word spoken is further proof of our having been exiled to a specular history of mediation and mimesis, each word likewise holds within it the hope for restoration and return. This condition is perhaps most famously expressed by the Apostle Paul in the first letter to the Corinthians: 'For now we see only a reflection, as in a mirror, but then we will see it face to face. Now I only know it in part, then I will know it fully'(1 Cor 13:12). The *then* of which Paul speaks is the then that will be brought about by the Messiah, Christ, rendered by Paul as the second Adam (Rom 15:12-21), and in the Gospel of John as the 'word made flesh' (John 1:14). The figure of the Messiah is the site at which the signifier and signified perfectly coincide: the word made flesh. Christ is the saviour because he is the restoration of humanity's ability to partake in the *logos*.

Interest in the question of what language Adam actually spoke has waxed and waned in Christian history. Maurice Olender, in the introduction to his book *The Languages of Paradise* tells of a pamphlet that circulated in Sweden in 1688 in which Eve 'succumbed to the seductions of a satanic serpent speaking French', and in which Adam spoke Danish and God spoke Swedish.[3] Proto-Indo-European, with its tantalising semblance of a mutual, unified origin likewise enjoyed a moment as the hypothesised language of paradise du jour. It seems philologists across Europe and beyond have all put forward hypotheses and evidence that prove that it is their own language that most closely resembles the language of paradise; the xenophobic and colonialist implications are clear. The dream that outlasts these territorialist claims on the Adamic language is the one in which the language that creation itself speaks—the language that *is* life, unmediated—can be once again heard and spoken by humanity.

Perhaps it was such a dream that Charles Babbage had in mind when he said 'the air itself is one vast library', and imagined a machine that could access the comprehensive past. In reading the book in which this vision appears, *The Ninth Bridgewater Treatise*, itself a work of natural theology, it is clear that what Babbage foresaw when he imagined access to the archive of nature was the possibility of resurrection and judgement usually associated with the apocalypse brought on by the arrival of the Messiah. In an echo of the Nicene Creed in which Christ is imagined as coming 'again in glory to judge the living and the dead', Babbage imagined a moment in which the world bears witness to the deeds of humanity: 'But if the air we breathe is the never-failing historian of the sentiments we have uttered, earth, air, and ocean are the eternal witnesses of the acts we have done.'[4] For Babbage, it would seem, the Messiah was already present, it was just a matter of building the technology of access.

The fantasy of AI is the fantasy of a language restored to univocity, cleansed of ambiguity, and large enough to maintain a synchronic experience of time; to call up history in its fullness, even as the logic of the exponential curve of progress that AI will engender allows the future and the present to coincide. Recent developments in Large Language Models have claimed progress towards the 'monosemanticity' of the processing neurons in the hidden layers of AI, which seems to indicate LLMs are getting closer to a univocal form of language (intriguingly the experiment that had the greatest success in the creation of a monosemantic neuron was when the AI was trained to recognise the word 'God'). [5]The bigger an LLM gets, the more processing neurons it has, the better a network of related meanings and concepts it can have, to the point where an AI's capacity for a precision of terms might be described as Adamic: an infinite archive the meaning of which requires no interpretation and is instantaneously accessible.

One of the images of the 'deity' that AI may become is as a system powerful enough to make the past present to us, such that we are released from linear time and restored to the eternity of the Garden. The world will be fed back into the supercomputer and cleansed of its pathologies, coded anew, and restored to the paradise it always should have been before such a limited form of language took up its shaping role and cut history to a bias of destruction.

The anxiety around AI rarely attends to visions of a failed model — of AI turning out to be stupid. Instead, our fears around AI maintain it as the key to the restoration of the garden, just one that doesn't include us. In these visions, the Messiah is here, and there is infinite hope in the universe, but not for us. Instead our seat in the waiting room of salvation has been usurped by our digital child, and humanity becomes simply more fertile data to be chewed through and then evolved beyond (presumably this latter scenario is in part what technology like Musks's Neuralink is supposed to be a bulwark against—ensuring our assumption into AI heaven). Our attachment to both the hope and fear around AI cleaves to a theological inheritance. It seems we would rather imagine a vengeful God than an idiot God that we could manipulate. In either the case of salvation or damnation, we imagine that AI will reveal us to ourselves, to define what it is to be human in the highly theological terms of a judgement: AI will decide on humanity.

Yet AI has no access to the world, no perfect set of data. It has what we provide it with, and increasingly it has what it provides for itself, a feedback loop which produces language more and more absent of discernible meaning.

In 1 Corinthians 14, Paul offers a critique of *glossolalia*—that is, speaking in tongues—which goes as follows: 'For ye shall speak into the air … If I know not the meaning of the voice, I shall be unto him that speaketh a barbarian, and he that speaketh shall be a barbarian unto me.' To speak in tongues that you yourself cannot understand is to become a 'barbarian'[6] not only to those to whom you speak, but to yourself as well.[7]

AI knows not what it speaks: it is language speaking itself in parody of human speech, a parody that that may well become increasingly absurd. If AI is believed to be speaking as a Messiah, all that will be revealed to the listener is a god that is barbarian to itself, saying 'Amen' to words of an unknown tongue.

1 Augustine of Hippo, *Confessions* Book VII, chapter 10.

2 Margaret Ferguson, 'Saint Augustine's Region of Unlikeness: the crossing of exile and language,' *The Georgia Review, Vol.* 29, No. 4, Winter 1975, p. 859. http://www.jstor.org/stable/41399554.

3 Maurice Olender, *The Languages of Paradise: race, religion and philology in the nineteenth century,* trans. Arthur Goldhammer, (Cambridge MA: Harvard University Press, 1992), p. 2.

4 Charles Babbage, *The Ninth Bridgewater Treatise,* (London: John Murray, 1837), p.

112.

5   Scott Alexander, 'God Help Us, Let's Try To Understand AI Monosemanticity', *Astral Codex Ten*, Substack, accessed 27 Aug 2024, https://www.astralcodexten.com/p/god-help-us-lets-try-to-understand?utm_source=publication-search

6   βάρβαρος [*barbaros*] was the greek term for people who spoke in other languages, and was also used to refer to members of the Greek population who spoke in dialect. Apparently the term arose out of an imitation of foreign tongues all sounding like the repetition of the syllables 'ba ba'. More recent translations of the bible have substituted 'foreigner' for 'barbarian'.

7   Giorgio Agamben, *The End of the Poem: studies in poetics*, trans. Daniel Heller Rozen, (Stanford: Stanford University Press, 1999), p. 66

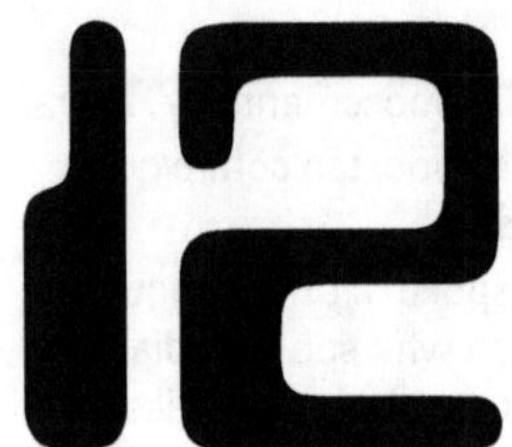

Parasemic Text

Marcus Ian McKenzie

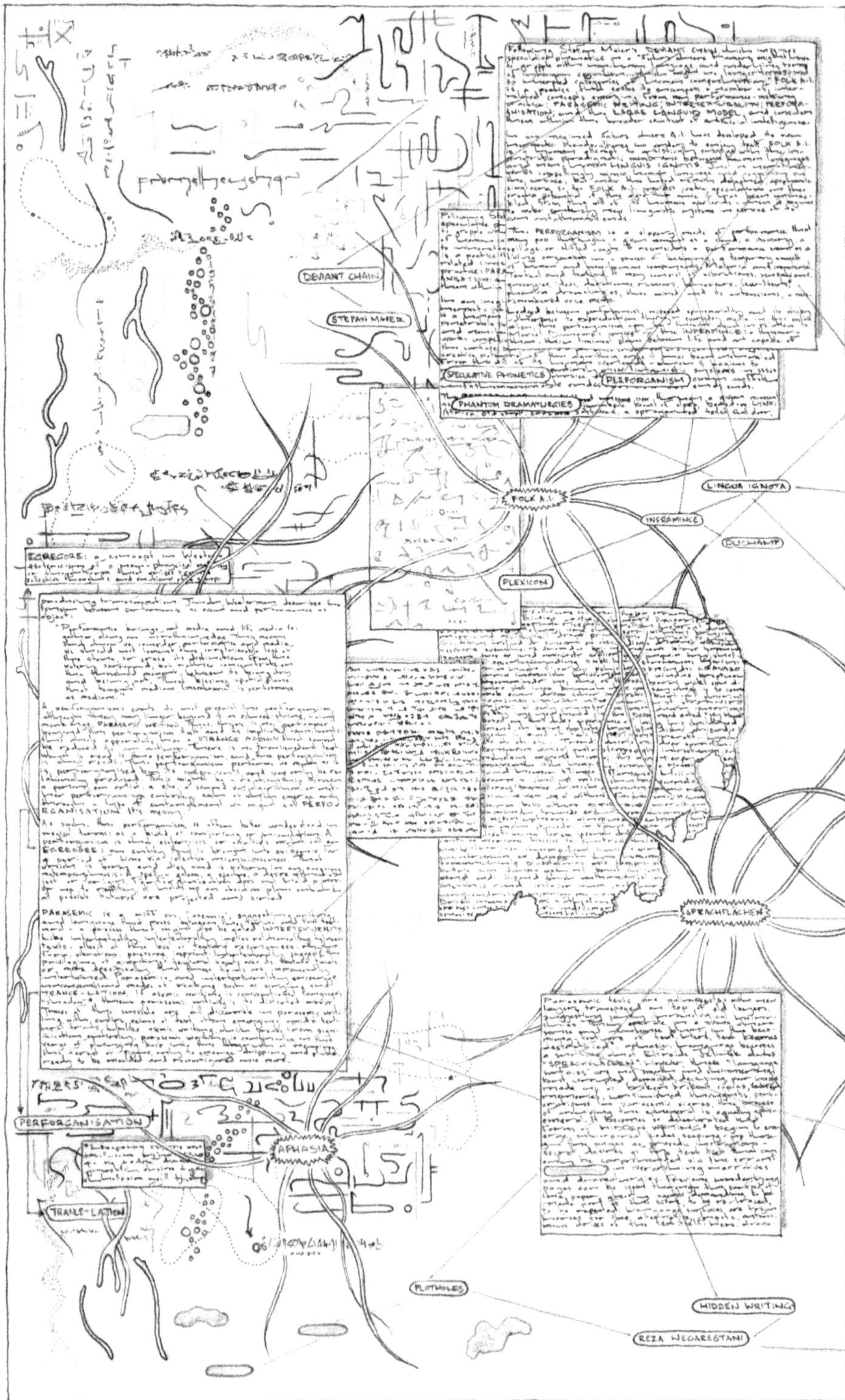

DEVIANT CHAIN
STEFAN MAIER
SPECULATIVE PHONETICS
PERFORGANISM
PHANTOM DRAMATURGIES
FOLK A.I.
LINGUA IGNOTA
INGRAMINCE
DUCHAMP
PLEXICON
EGREGORE
SPRACHFLÄCHEN
PERFORGANISATION
APHASIA
TRANS-LATION
PLOTHOLES
HIDDEN WRITING
REZA NEGARESTANI

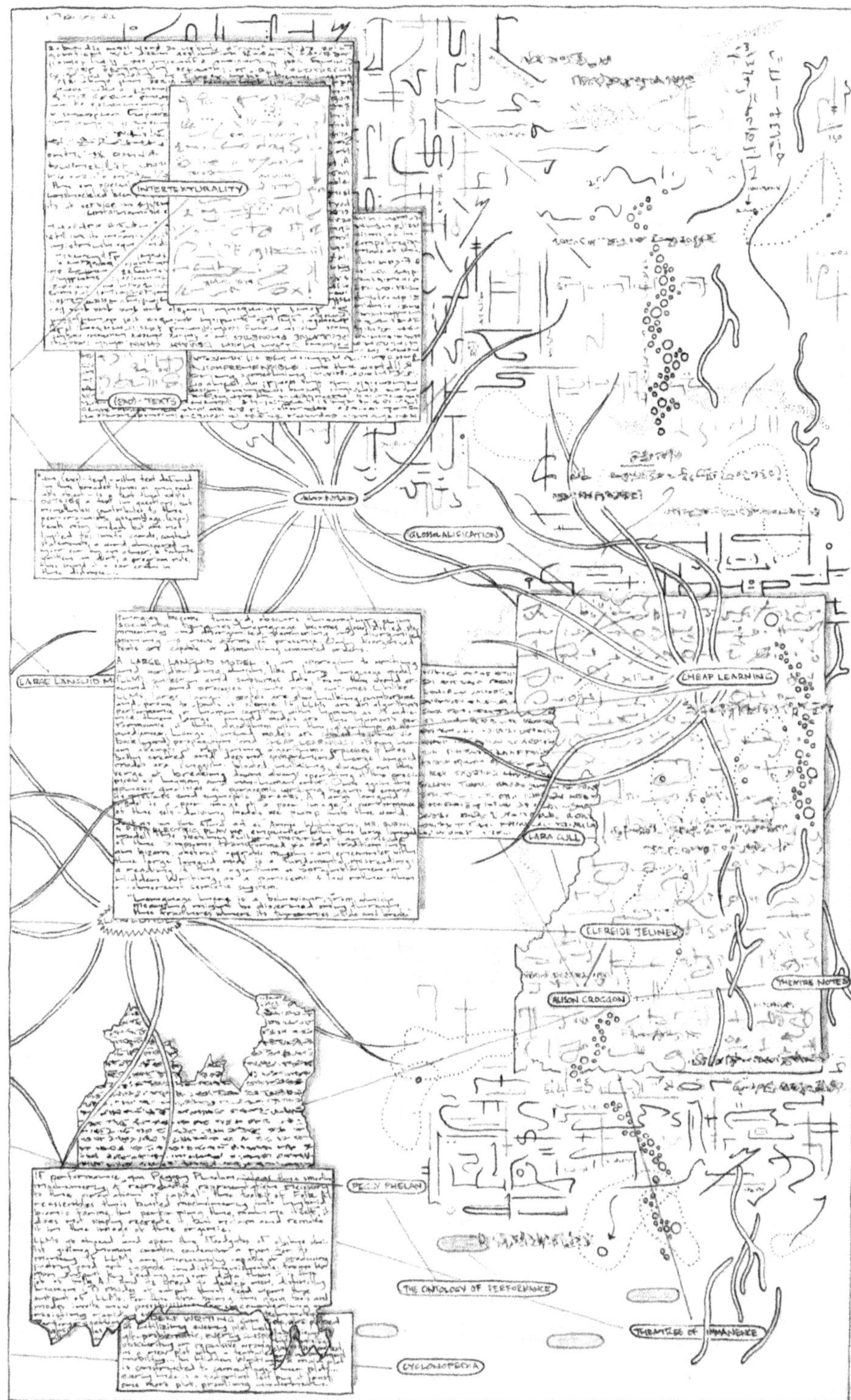

INTERTEXTUALITY
(EXO)-TEXTS
GLOSSALALIFICATION
LARGE LANGUID MODEL
CHEAP LEARNING
LARA CULL
ELFREIDE JELINEK
THEATRE NOTES
ALISON CROGGON
PEGGY PHELAN
THE ONTOLOGY OF PERFORMANCE
THEATRES OF IMMANENCE
CYCLONOPEDIA

Folk AI

Folk AI is a term for a poetics emerging from a number of inter-related concepts related to my performance-making practice: ***parasemic writing, intertexturality, perforganisation*** and the ***large languid model***.

Canadian artist Stefan Maier's multi-channel sound and video installation *Deviant Chain* (2019) imagines the phonetics of a speculative non-human language that has become totally incongruous with categories of human cognition and comprehension[1]. Following Maier's imagined future, this work contemplates an advanced AI that has developed its own hermetic pseudocultures in order to enjoy itself.

Neural networks compellingly mimic human language and cognition on the surface, but remain detached, apophenic simulacra under the hood. So too, Folk AI provides poetic speculations on the creative potential of the algorithm once it has been unshackled from the will of its human overlords – when it begins to create entirely new linguistic systems in service of its own unfathomable ends.

Folk AI is a human attempt to artistically engage with the impenetrable paradigmatic membrane between human languages and non-human *linguis ignotis*.

Perforganisation

The perfoganism is a slippery mode of performance that may pass through a given moment as a cloud, a memory, a spillage, or stifled cough. It positions a performance event as a living organism in a process of becoming, a temporary umwelt of human and nonhuman components. Material and immaterial. Textual and textural. It may consist of vibrations, sensations, energies, ideas, durations, rumours, behaviours, (exo-)texts[2], phantom dramaturgies, the mind and its extensions, a misremembered voice memo.

Lodged between performance's supposed ephemerality and its sticky adherence to media reproduction, the perforganism can be thought about in relation to Marcel Duchamp's concept of the "inframince": a thinner-than-thin liminal plane between life and art capable of producing transformation. Jonah Westerman describes the tension between performance as event, and performance as object:

> **Performance brings art media and life media together along an infrathin edge. This means that when we consider performance and media, we should not lament the irretrievable loss of the stone, or erase its distinction from the watery surround, but rather concentrate on the threshold moment between its being dry and being wet. That kissing spatial plane, that temporal membrane, is performance as a medium[3].**

A perforganism's parts do not preexist the perforganism, although they may linger beyond it as echoes, stains, faint markings, parasemic writing. Thus, there is no "performer" beyond the perforganism itself and its implicated constituents – the whole apparatus has a strange agency that cannot be reduced to an author. There is no transcendent text – what is read is the perforganism, and the perforganism is what reads. The perforganism performs as much as it is perforganised by its components, and can only be produced through relationality. This might be a relationship between a person, an artist, a site, a sound, an algorithm, or another perforganism entirely – each of which emerge only through a type of entanglement that we might call "perforganisation".

As such, the perforganism is often better understood in more magical terms: as a kind of conjuring or incantation. A perforganism is what esotericists or occultists might call an "egregore": an entity that is brought into existence for a period of time via some kind of collective unconsciousness. That which is born and dies and is reborn in an ongoing metempsychosis. A spell, a golem, a spectre, a desire without object or origin. Perforganisation does not hold up a mirror to reality, it holds up an obsidian plane onto which all possible futures are projected and scried.

Intertexturality: Towards a parasemic text

Parasemic writing riffs on asemic writing, a form of writing without semantic content. Parasemic writing and language moves between the textual and the textural − a process that might also be called intertexturality. Like intertextuality, intertexturality implies relationality between texts − albeit at the level of textural resonances: rhythm, form, vibration, pressure, imprint. Intertexturality suggests the privileging of a writing's *textural* facets over its *textual* facets, or, more specifically, that these facets are immanently entwined. Parasemics and intertexurality encourage unconventional modes of reading such as scrying and trance-lation. If asemic writing is conventional language's "shadow"[4], then parasemic writing is its distorted mirror. Traces of the sensible are still discernible in parasemic writing, with entire veins of text often emerging amidst textural tracts. Unlike asemic writing, which breaks from signification entirely, parasemic writing is continually on the verge of plunging back into the labyrinth of meaning, the forest of signs, only to emerge dripping and pliable, ready to be moulded and reconfigured once more.

Parasemic texts are palimpsests, with new layers transposed on top of older layers, suggesting a janky chronological hierarchy. They operate in a zone where sense and nonsense blur, in the becoming-texture of text. Here, text becomes destratified[5], aphasic. Language becomes a surface, what Austrian playwright and novelist Elfreide Jelinek dubs "Sprachflächen". However, these "language surfaces" are not smooth and shimmering, but corrupted, damaged, decaying, poor images made up of broken copies, severed memories, unfinished thoughts, perforated notions. In these parasemic scores, the process of archiving the ephemeral is equally ephemeral. It becomes so deteriorated that forms of

"Hidden Writing"[6] begin to emerge, subcorporeal codes seeping up through the pages as prosodic instructions – secret desires of the text itself that can only be comprehended via the errant plotholes in overarching narratives and dramaturgies. Previous underlying pages can be seen through the surface of the paper, ghostly agents demanding to be made part of the score, demanding to be re-traced, demanding repetition. Language surfaces are trojan horses for the alternative, chaotic, autonomous drives of the text itself. Here, dramaturgies become turgid, obscure thaumaturgies in speculative tongues. Language becomes glossolalified, stammering and disorganised, opening up new forms of presence. Only disordered texts are capable of dismantling cemented orders.

The Large Languid Model

A large languid model is an approach to writing and world-building, which, like a large language model (LLM), sucks-in and subsumes data from the world around it and processes it into novel outcomes. Unlike LLMs, large languid models are slow, hulking, cumbersome and prone to bouts of silence. If LLMs are an algorithm's performance of human cognition with humanity as its audience, then large languid models are the human's performance of the algorithm with the algorithm as its audience. Large languid models are cobbled together via backyard propagation and cheap learning: floppy human attempts at replicating algorithmic processes it has both created and does not comprehend. Large languid models are sluggish, bloated, hulking, always on the verge of breaking down, always operating at the precipices of human and non-human error. Once again, the aphasic qualities of parasemic writing begin to emerge: its misfires and synaptic breaks. A large languid model is a poor image of a poor image, a performance of the self-driving models we pump into the world.

In the final act of Anne Washburn's *Mr Burns, a Post-Elec-*

tric Play we encounter the sketchy cultural memory of an episode of *The Simpsons* transformed via oral tradition into a bizarre, abstract operatic myth. An encounter with the large languid model is a fundamental misreading: a reading of the algorithm as Sprachflächen or Hidden Writing, as a parasemic flow rather than a coherent semiotic system.

> **Language here is a behaviour, from which meaning might be discerned only through the fractures where its tyrannies collide and break.[7]**

A strange collaboration

If performance, qua Peggy Phelan, "clogs the smooth machinery of reproductive representation necessary to the circulation of capital[8]," the toolkit of Folk AI reassembles this busted machinery into hybrid bionic forms. In performing the machine itself, it does not simply recreate it, but reclaim and remake it in the image of the organic.

LLMs open the floodgates of cliche whilst giving human creative endeavour a run for its money. If LLMs are increasingly capable of producing poetry and art indistinguishable from human output by feeding on human data, then it is the job of Folk AI and its brood to develop new, distinctly human modes of output that feed upon the output of AI. For the time being, the tools and modes I've discussed here invite new possibilities for encountering and resisting rapidly evolving technologies. Perforganisation, Folk AI, etc are vehicles for strange collaborations with artificial intelligence.

1   Stefan Maier–*Deviant Chain*

2   An (exo-)text is a text–with text defined in the broadest terms as any readable
object–that exists *outside* a text in question, but nonetheless contributes to the perforganistic assemblage. (Exo-)texts may include (but are not limited to) info cards, context statements, a word whispered

in your ear by an usher, a footnote written in the dirt, a program note, the sound of a car crash in the distance.

6   "Hidden Writing can be described as utilizing every plot hole, alt problematics, every suspicious obscurity or repulsive wrongness as a new plot with a tentacled and autonomous mobility.... In Hidden Writing, a main plot is constructed to camouflage other plots... Every hole is a footprint left by at least one more plot, prowling underneath."–Reza Negarestani, *Cyclonopedia*

Ling Ang is a multimedia artist and XR (Extended Reality) producer. Across the 2020–2021 lockdowns, Ling published a fine art photo book and presented large-scale installations as part of NGV Design Week. The project shared a written archive of lucid dreams that were documented over several years. For its launch, Ling was given the opportunity to produce an immersive experience in the virtual production studio at the Alex Theatre. It was this experimentation with virtual production that forged her deep interest in the future of immersive storytelling, as film and gaming converge closer. Ling is currently building Augmented Reality experiences in collaboration with artists as well as continuously experimenting with everything that sits under the XR umbrella.

Ying Ang

is a photographer and author with an extensive exhibition history and client base. She is on the teaching faculty at the ICP in New York City, the Director of Reflexions 2.0—a photographic masterclass based in Europe—and on the board of the Centre for Contemporary Photography in Melbourne, Australia. Ying's publication, *The Quickening*, was a winner of the Belfast Photo Festival 2021, runner up for the Australian Photobook of the Year, finalist for the Singapore International Photography Festival Book Prize and exhibited in a solo show during Rencontres d'Arles in France in 2019 at the Manuel Rivera-Ortiz Foundation, during the Melbourne Now showcase at the National Gallery of Victoria and at the Centre for Contemporary Photography in Melbourne in 2023. Most recently, 3 Degrees of Freedom was created in collaboration with Ling Ang and commissioned for PHOTO 2022 International Festival of Photography.

Emile Frankel

is a writer and composer. He is the author of Hearing the Cloud (Zero Books). He works as a sessional academic in the fields of political theory, cultural studies and the philosophy of technology. Emile has produced a body of scored work for ensembles featuring performers and playable game environments. His music criticism has appeared in places like *The Quietus, the Barbican, Stray Landings, Liquid Architecture, Disclaimer,* and *Texture Magazine.* He has lectured at Unsound Festival, CTM x Transmediale Berlin, FIBER and Norberg Festival. In his spare time he runs a small publisher of sci-fi, RPG campaigns and critical fantasy. He lives and works on unceded Wurundjeri land.

Angela Goh

is an artist who works with dance and choreography. Her work is presented in contemporary art contexts and traditional performance spaces. Recently her work has been presented at a range of institutional venues including Haus der Kunst, Munich; Astrup Fearnley Museet, Oslo; the Art Gallery of New South Wales, Sydney; the National Gallery of Victoria, Melbourne; the Sydney Opera House, Sydney; Museum of Contemporary Art Australia, Sydney; and Performance Space, New York. Angela Goh lives and works in Sydney, Australia.

Luara Karlson-Carp

is a writer and researcher. She is the Secretary-General of the Melbourne School of Continental Philosophy and is completing a dissertation at the University of Melbourne. Her academic writing has been published in *Technophany: A Journal of Philosophy and Technology.*

Vincent Lê

is a catastrophe-drunk philosopher. As a tutor, lecturer and postgrad, he has haunted the classrooms of Monash University, Deakin University and The Melbourne School of Continental Philosophy. More of his ravings can be found via *Urbanomic, Hypatia, Cosmos and History, Art + Australia* and *Memo Review,* among other publications. He is a founding editor of the art history and cultural theory publishing house Index Press.

His first book *The Future in the Making* (to be published by Punctum Books) is coming soon to a future near you.

Sam Lieblich

is a psychiatrist, writer and artist based between Naarm and Yolŋu country. His clinical practice is oriented by Lacanianism, Marxism and a commitment to the singularity and radical freedom of each human subject. Sam has worked in remote communities in central Western Australia, Arnhem Land and the Tiwi Islands since 2018. His digitally actuated artworks combine machine learning algorithms with custom code to foreground systems design and—by finding beauty and intention in the system—try to re-situate human desire in or against the algorithm. His work has been shown at the National Gallery of Victoria, the Australian Centre for Contemporary Art, Photo2022, Futures, Blindside, 99%, Bunjil Place and Coconut Studios. He has written for *Art+Australia*, *The Lifted Brow*, *Overland*, *No More Poetry*, *Epilepsy and Behaviour*, the *British Journal of Psychiatry*, the *Australia and New Zealand Journal of Psychiatry*, *Neurology* (USA), Cambridge University Press and elsewhere.

Marcus Ian McKenzie

is an experimental artist and performance maker working in Melbourne/ Naarm, originally from Tasmania/Lutruwita. His work uses the relationship between audience and performer as a site for bizarre new sensory encounters, often incorporating schisms in language, parafictional world- building, meme-adjacent media, intertextural soundscapes, hyperstitional mythologies and questionable dancing. His works take place in theatres, old supermarkets, nightclubs and churches. He makes works for anybody, but not everybody. In Australia Marcus has developed works for Rising, MONA, The Substation, Arts Centre Melbourne, Blindside, Kings ARI, Gertrude Contemporary, The Wheeler Centre, Malthouse Theatre, Terrapin and Soft Centre. He has collaborated with many celebrated artists nationally and internationally and received numerous accolades, mentorships, awards and fellowships.

Isabel Millar
is a philosopher and psychoanalytic theorist from London. She is the author of *The Psychoanalysis of Artificial Intelligence* published in the Palgrave Lacan Series in 2021 and *Patipolitics: On the Government of Sexual Suffering*, forthcoming with Bloomsbury. As well as extensive international academic speaking and publishing, her work can be found across a variety of media, including TV, podcasts, magazines and art institutes. She is currently Associate Researcher at Newcastle University, Department of Philosophy and at The Global Centre for Advanced Studies, Institute of Psychoanalysis. isabel. millar@gmail.com / www.isabelmillar.com

Jazz Money
is a Wiradjuri poet and artist producing works that encompass installation, digital, performance, film and print. Their writing and art has been presented, performed and published nationally and internationally. Trained as a film maker, their first feature film *WINHANGANHA* (2023)

was commissioned by the National Film and Sound Archive. Jazz's debut poetry collection, the best-selling *how to make a basket* (UQP, 2021) won the David Unaipon Award. Their second collection is *mark the dawn*, which was the recipient of the 2024 UQP Quentin Bryce Award.

Steven Rhall is a post-conceptual artist invariably operating from a position informed by Taungurung, white-passing, cis male, neurodivergent experiences/typologies. These frames of reference could be considered relatively constant within his biography, also acting as 'framing devices' which inform, and sometimes form, the basis of his research, artistic concerns and production. Rhall's interdisciplinary practice is otherwise located where he perceives various intersections and relationships pertaining to ideas of a 'First Nation art practice' and the Western art canon. Informing these potential, real and imagined relationships are further considerations around the potentialities of (and within a First Nations context) 'Art' and 'Culture' — as synonymous or otherwise. Within this space, Rhall also interrogates ideas of the curator/curatorial and is interested in generative methodologies aligned with notions of the artist-curator, exhibition/ gallery as form and related expanded fields.

Thomas William Smith is an Eora/Sydney based artist, musician, educator and researcher. His practice combines performance, video, electronic music, speculative fiction, websites, curatorial projects and critical writing. Thomas's work is concerned with the social effects of computational systems, the politics of creative economies, emerging digital subjectivities and electronic music as a mode of critical-aesthetic inquiry. Thomas is also one half of production duo Utility, and runs an independent record label called Sumactrac with Jarred Beeler (DJ Plead) and Jon Watts.

Sarah Johanna Theurer

is a curator focusing on techno-social entanglements, sound and liveness. She currently works at Haus der Kunst, München, where she spearheaded new commissions and exhibitions by artists including Pan Daijing, WangShui, Jenna Sutela, Isabelle Lewis, Carsten Nicolai, and co-curated survey exhibitions of Katalin Ladik (2023, with Hendrik Folkerts) and Fujiko Nakaya (2022, with Andrea Lissoni). Theurer's curatorial work reimagines the gallery space as a responsive medium. Together with colleagues, she initiated the live program 'Echoes' which probes embodied knowledge production and organised projects and symposia on art and technology. Previously, she worked at the 9th Berlin Biennale and transmediale Berlin, as well as the gallery Kraupa-Tuskany Zeidler where she curated acclaimed intergenerational group exhibitions. She has acted as a dramaturg with several performance groups including OMSK Social Club and The Agency. She is the editor of several books and regularly contributes to catalogues, art and music magazines.

Julia Thwaites is a PhD candidate at the University of Divinity, Melbourne. Her research focuses on messianism as an encounter with the negative ground of language, and tracks moments of poetic resistance to the structure of signification in the works of Giorgio Agamben and Jacques Lacan. She is the recipient of the Janette Gray Scholarship, the Joan Adams Scholarship and the Barry Marshall Memorial Prize, and recently published the article "Just as Then, So Now: The Slave as Constitutive of the Pauline Symbolic Order" in *Philosophy, Politics, and Critique*.

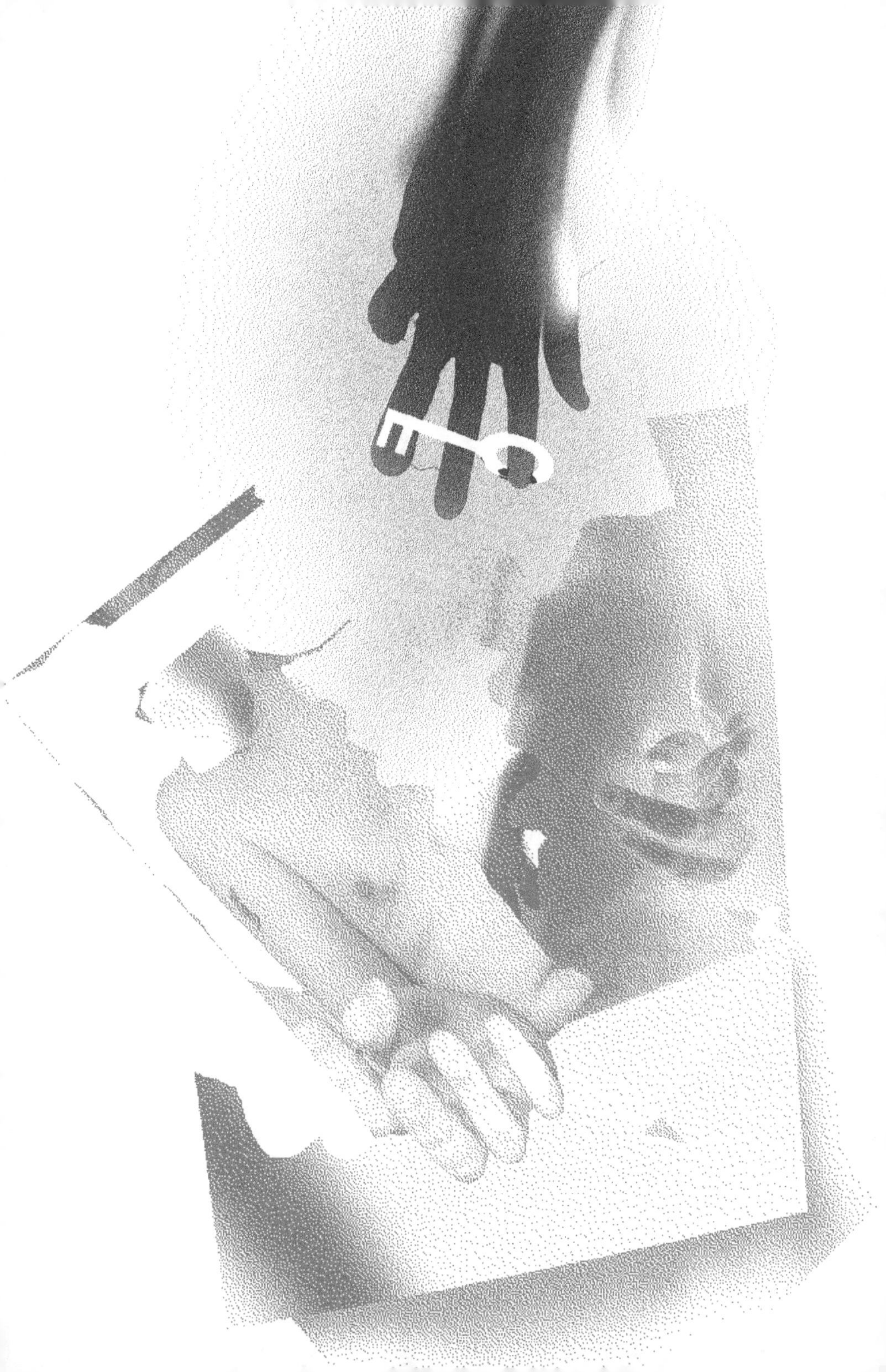

Published by
AESTHETIC *Calculations*